Remote *Drone* Pilot Cer[tifica...]
Study Guide

"Your Key to Earning FAA Part 107 Remote Pilot Certification"

Guaranteed to Pass or Your Money Back! *

For Commercial, Government and Law Enforcement Agencies

RemotePilotAssociation.com
TC FREEMAN

WOF Media

www.RemotePilotAssociation.com

Request for information should be addressed to:

WOF Media
Attn: TC Freeman
747 S. 8th Street
Mebane, NC 27302

Cover art by Lauria

Edited by Proof Reader Pro 5

Book formatting by Dumitrup

Developed and produced in the United States of America

About the Author

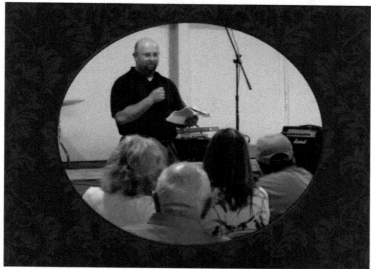

TC FREEMAN, RemotePilotAssociation.com, Aviation Safety Specialist

TC Freeman was raised in an aviation family handling the controls of an aircraft before he could qualify for a driver's license. After High School he attended Guilford Technical Community College (GTCC) earning an A.S. Career Pilot/Aviation Management. Shortly after graduation, he received a Flight Instructor and Instrument Instructor certificate as well as a Multi-Engine Rating. He taught flying and eventually went back to GTCC gaining another A.S. in Aircraft Systems Technology (FAA Airframe and Powerplant certificate). In 2013 he completed the goal of attaining a B.A.S. Degree from Campbell University, Research Triangle Park (RTP), North Carolina.

For the past 20 years TC has been employed as an Aviation Safety Specialist for the government presenting education and safety programs to pilots, Aviation Maintenance Technicians (AMTs) and the general public. He served as a pilot on the Cessna 441 Conquest aircraft which lead to the attainment of an Airline Transport Pilot certificate.

TC serves on the Advisory Board of the Wayne Community College Aviation Systems Technology program, NC Aviation Museum (Asheboro, NC), Elizabeth City State University (ECSU) Aerospace Program and the NC Unmanned Aerial Systems (UAS) Safety Group.

TC's first eBook, "How To Save Money On Flight Training" was published in 2012 followed by audio presentations, "Operations at Non-Towered Airports" (2013), "Pilot and Aircraft Performance" (2014) and "Flight 5491 Unraveled" (2016). Presently he is working on additional new material that will benefit drone (UAS) pilots.

Contact TC FREEMAN at tc@tcfreeman.com or call 919-619-6828.

Dedication

"To my Father, Mother and Step-Mother, thanks for your love and support. I couldn't think of a better upbringing than being taught to fly by my Father. Also, my daughters, Hope and Jessica, I'm so glad you turned out to be inspiring young women. Special appreciation to Dr. Phil Deaton and Von Miller, I wouldn't be in aviation without your mentorship."

Table of Contents

Disclaimer

The drone (UAS) industry and the regulations are rapidly changing and adapting to the nature of this new commercial opportunity. As of this writing references to the regulations and related material are current, but we encourage you to research the most current information at www.faa.gov/uas.

The study guide questions are a good representation of what will be on the test based on our research. Some questions may be exact, while others are similar. There is a chance that we will have omitted some questions we felt were not applicable to the Remote Pilot certification but the FAA decided to ultimately include. We offer a money back guarantee if you do not pass the Knowledge Test. Simply send us a copy of your test results and we will refund the price of this book, no questions asked.

Please help us make this study guide the best possible by giving us feedback about your experience with the material and Knowledge Test (and your passing grade). Our passion is to see a lot of good folks; start a small business, use a drone in their company or government organization, to become a safe and successful Remote Pilot.

Chapter 1

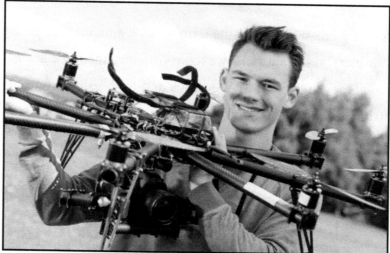

Source: CanStockPhoto

Setting the Table

Introduction

This is an exciting time in the budding commercial drone pilot industry. As of this writing (July 2016) the Federal Aviation Administration (FAA) created an easier path for commercial flight operations. Drone pilots, with and without existing FAA certification, will be able to attain a certificate for Remote Pilot commercial flight on August 29, 2016. **Testing centers** across the United States that handle FAA computer testing, called Knowledge Test, are getting ready to put the FAA Remote Pilot Knowledge Test on their computers. Quite frankly, the transition has been quite confusing to many, especially those that are new to the way the FAA does business. We are here to help you pass the FAA Remote Pilot certification course and get real knowledge that can be used in the field.

Prior to the birth of the Remote Pilot (RP) certification, drone pilots that wanted to fly commercially had to submit a letter, called a 333 Exemption, to be considered for approval. The process was extremely slow and left many waiting by the mailbox for an FAA approval letter. The good news is that the new process is here and we highly suggest taking the Knowledge Test verses waiting on a 333 Exemption. Your mailman will thank you. But first, let us clear the air on terminology used in this study guide.

Some industry/FAA officials want to use the term small Unmanned Aerial System (sUAS) instead of drones due to the military association. However, most consumers are accustomed to the term drone, the name under which they are sold. I may go back and forth using drones, Unmanned Aircraft (UA) and sUAS all referring to the consumer/prosumer type drones that weigh less than 55 lbs.

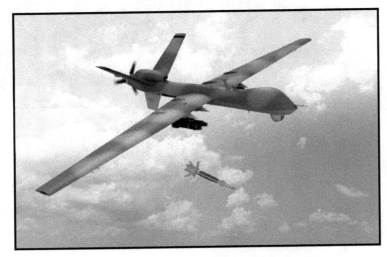

Military Drone, Source: CanStockPhoto

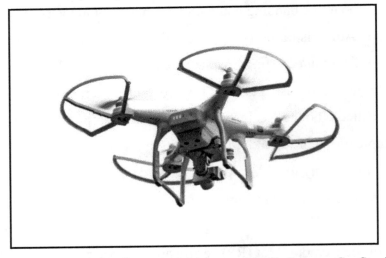

Consumer/Prosumer Unmanned Aerial System (UAS), Source: CanStockPhoto

I personally get concerned about the great use of acronyms in aviation causing fewer people to be interested in the industry (and hobby) because they are intimidated by the complex terminology. As a small example, the FAA calls the flight and ground test for earning a pilot license an Oral and Practical Test (O & P). Most pilots historically simply called this final test a "check-ride," which seems to fit the need quite well in my opinion. By the way, the Part 107 commercial Remote Pilot certification computerized test is called a Knowledge Test (KT). There is a certain amount of terminology is important to know as a baseline of knowledge and we will integrate this into the course. We have also included a glossary if you forget an acronym.

After taking a massive amount of FAA test (approximately 10) I realize there are various ways in which to teach the material. I like to good balance of getting to the point material and short cut's (hacks, in a good learning way) mixed in with personal commentary and a dash of humor. As in life, sometimes you have to understand the background of a topic to fully get the message, otherwise you are just memorizing. We want to build a solid foundation of knowledge.

Changes

If you are concerned about changes to this material, don't worry, we have you covered. Any updates, changes or clarifications will be found at www.RemotePilotAssociation.com. I'll bet you a *DJI Inspire* drone that the FAA will change the regulations covered under Part 107 as issues arise. It's important to stay on top of the latest information regarding regulation because your certification may depend on it.

Not Your Typical Study Guide

If you are looking for your dry, exhausting and, quite frankly, boring study guide, you have come to the wrong place. Our objective, I like to say our secret sauce, is to give you the material you need to pass the test in a fast, fun and efficient format. We also won't leave you hanging and will be there to answer questions along the way on our website; www.RemotePilotAssociation.com (RPA). This book comes with 1 FREE year Membership with a bunch of perks (look for our email confirmation), but before I start sounding like a commercial let us get to the business at hand.

Links and Graphics

If we reference a website or graphic with a long link so we will post at;
www.RemotePilotAssociation.com/links.

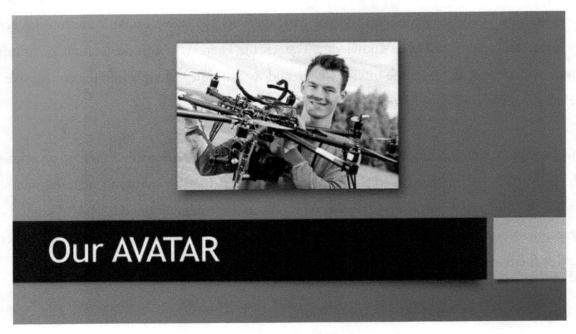

Source: CanStockPhoto

When developing this study guide we had a typical Avatar, or person, in mind. Our Avatar owns a drone (or more), a DJI Phantom 3 for example, and would simply like to start a small business, use in a company, or in a governmental organization. We are particularly inspired by the great opportunities available to an entrepreneur that start a commercial drone business with minimal investment. The good news is that any type of Small UAS (sUAS) operator can use this material to pass the FAA Remote Pilot Knowledge Test.

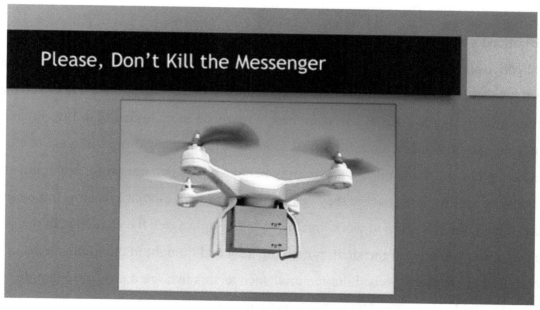

Source: CanStockPhoto

Some of the information covered is controversial, for example, the use of First Person View (FPV) devices. We have to cover some material, while we may not personally agree it is mandated by the FAA. Sometimes we will voice our opinion and we will be clear and obvious. The bottom line is that we are proponents of the industry, but we have to cover topics that are controversial in nature.

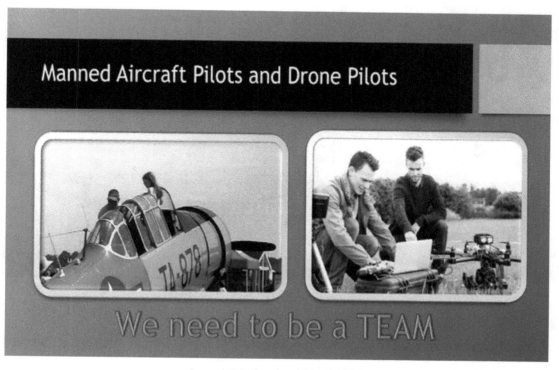

Source: RPA and CanStockPhoto

I have noticed a budding rivalry online and at industry gatherings about the introduction of drones into our collective airspace, or the National Airspace System (NAS) as the FAA calls it. I just read an article about the spike of drone sightings reported by manned aircraft since the FAA established its formal reporting system. While I see the concern, early reports do seem higher than I would expect. Perhaps this indicates some manned aircraft pilots discontent with infusing drones into the NAS. On the other hand, there plenty of drone pilots that strongly believe there is no reason for any regulatory oversight and should be able to fly anywhere they want. From my vantage-point I feel it is important for manned and unmanned aircraft pilots to work together. Most people agree the challenge to the sUAS industry is the customer who buy off the shelf, not giving a second thought to the implications of safely flying in the NAS. People that searched out material such as this that are a positive example of responsible and professional UAS flying. The technological advancements in the drone industry will bring about advances in manned flight. I'm afraid that the manned pilots that perceive drones as a threat will get left behind by not embracing this formidable budding industry. The bottom line is that we are ultimately stronger together in the aviation community than separate.

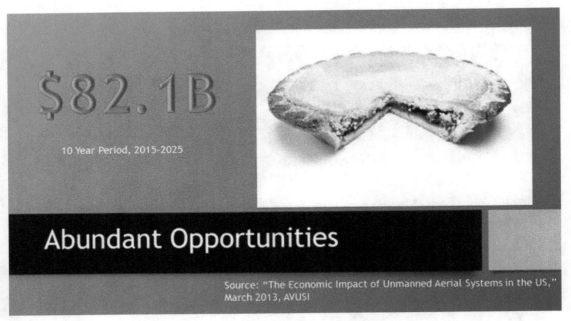

Source: RPA and CanStockPhoto

Speaking of rivalries, I want to take the opportunity to address the fact that some people reading this material could perceive fellow course takers as competitors rather than teammates. Industry estimates show that the UAS industry can mean big business, to the tune of $82.1 Billion between 2015 and 2025. As large corporations, I don't have to name-names here, reach for a considerable portion of the pie it is important for small drone business owners to band together. If we begin to back stab each other in the name of competition the big corporations will only become a more powerful lobby to Washington, DC. I'm not naïve enough to believe competition doesn't exist, but please consider the power of banding together in-light of the massive opportunities.

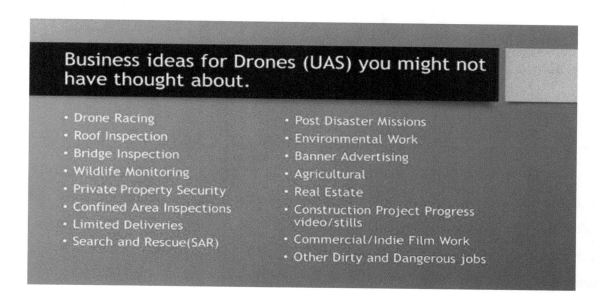

Note: Limited deliveries are mentioned in the above list, however the exceptional governmental hoops (waivers, etc.) you have to go through make it impractical. The Part 107 regulations (more on this later) are riddled with phrasing basically stating, "no deliveries by drone." I like to call this the "No Bezos Rule."

Many people are familiar with the obvious uses for drones, but there may be a few on this list you have not thought of and new applications are being developed every day. Personally, I'm very excited about Search and Rescue (SAR) and the task that reduces or eliminates the hazards to safety (dangerous and dirty jobs). Drones can be sent into burning buildings, confined spaces and more.

The other area with great potential is in entertainment, specifically drone racing. While there are some good examples of this sport in limited locations, and via video, I can see an opportunity to improve the value for the fan by increasing viewability and pilot personalities.

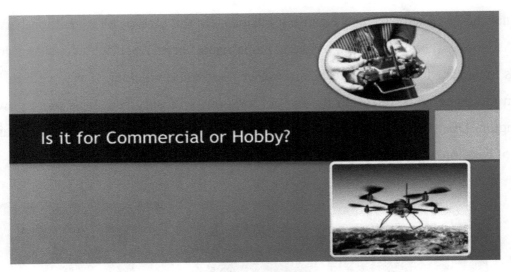

Source: CanStockPhoto

One question people have is what constitutes a hobby from commercial drone flying. While the answer would seem straight forward the difference can be more subtle in some areas. In its most basic definition commercial drone flying is a matter of doing the flying for compensation (be it direct money, barter or trade). Hobbyist flying is only for fun and while the product can be the same, like video of a scenic river or aerial views of a historic home, flying can more likely become a commercial endeavor. Some rouge pilots intentionally look for the gray areas in which to operate, but if you want to be in the business for a long time we don't suggest this route.

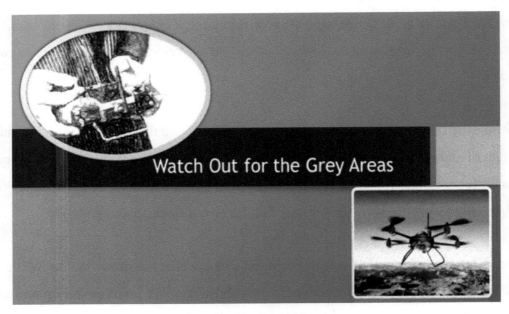

Source: CanStockPhoto

One seemingly harmless activity is a hobbyist posting a drone video on a video content website, such as YouTube. If the drone pilot simply checks the box for placing ads on the video then the work becomes commercial. Other examples are not as subtle, but more than likely a "work-around" such as, not charging a company for video footage or only charging for editing services but not the flying. Unfortunately, this falls into the commercial because the ultimate use of the video is for commercial purposes.

One situation that arises is, for example, the car lot owner or real estate agent that buys a drone to take video of the car lot, car or homes for sale. Buying and operating the drone as a business owner doesn't exempt the pilot from being deemed a commercial operation. Of course hiring or asking someone else to do the video is also commercial.

As you start your flying you will undoubtedly run into pilots unknowingly violating regulation or simply are not aware they are operating illegally. For the sake of the industry safety and fairness, encourage these folks to take the Remote Pilot Knowledge Test (RP KT). It's our responsibility to take reasonable steps to stop someone from "careless and reckless" (covered under regulation Part 107.23). Known commercial operators that are not licensed can approached and/or reported to the FAA. People report illegal aircraft charters and illegal commercial drone flying is no different in my opinion. However, I understand some of the dis-satisfaction some people have with the regulations that have been put in place by the county, city, state and federal government. A mass of laws from multiple sources (although the FAA is working to get a handle on this) means that well meaning pilots can unknowingly be in violation. Our recommendation is to band together to tackle issues with regulation in a formal manner. If you have a question as to if a flight is commercial or hobbyist contact us directly at; www.remotepilotassociation.com.

If you want to pass along good information to hobbyist pilots send them to be sure to check out the Advisory Circular titled, "Model Aircraft Operating Standards," at; http://www.faa.gov/documentLibrary/media/Advisory_Circular/AC_91-57A_Ch_1.pdf.

Who Really Controls the Skys?

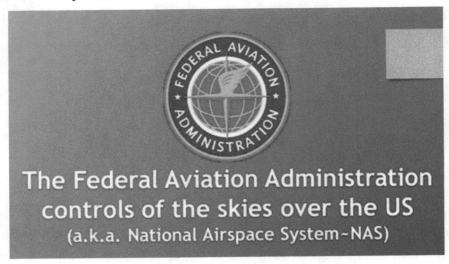

Logo Source: FAA

The FAA has made it very clear they are in control of the sky in which we fly. There are many examples of videos on YouTube, and the FAA has used "enforcement action" against some of these UAS pilots. We like to call this FAA action, "enforcement by YouTube." While we are not in a position to make legal calls, federal law typically trumps State, county, city and private property rules. However, many of these groups still might use the legal system to restrict drone use. The FAA has attempted to encourage States to allow federal laws to govern UAS. While time-consuming, we recommend practicing "due-diligence" with respect to obtaining various permissions from these organizations.

This Point Forward

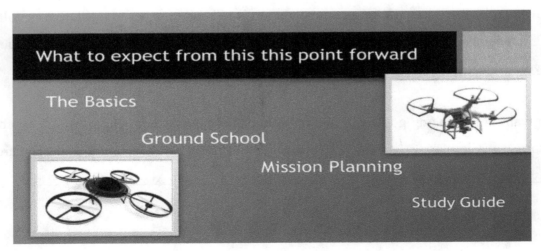

Source: CanStockPhoto

We are going to jump right into earning your Part 107 Remote Pilot Certificate but wanted to cover a quick bit of history on gaining FAA commercial drone pilot authorization. Early in the history pilots (typically companies) Applied, called a petition letter, for an FAA 333 Exemption. What is an FAA 333 Exemption? Essentially, it is the process of asking the FAA to release a drone pilot (or company) from select FAA regulations that primarily applied to manned aircraft. This issue is now resolved with the new Part 107 regulations, the reason you are using this study guide. However, you can go through a similar process to apply for a waiver to allow for exemption from regulation in flight operations such as; night flying, flying over 400' Above Ground Level (AGL), flight over people and more. Waiverable items are covered under Code of Federal Regulation (CFR) Part 107.205 (more on this later).

Pilots that already have FAA certification (over a Student Pilot Certificate), such as a Sport or Private Pilot with a current Flight Review can take a Part 107 online course at www.faasafety.gov. After completing the online test these pilots will fill out an application on IACRA (online form system the FAA uses) to complete the process of getting Remote Pilot certification added to their credentials.

A sUAS pilot without existing FAA certification will be able to take the Part 107 Knowledge test at testing centers as of August 29, 2016.

In addition to the basics, we will cover general "ground school" information that is applicable to sUAS. Ground School is the term used by the majority of manned aircraft pilots referring to the textbook (or coursework) required to complete a Knowledge Test. This will be followed the study guide portion of the material, with some practical mission planning tips along the way. The study guide will consist of representative questions for the Remote Pilot Knowledge Test in which you can answer then scroll down to see the answer with an explanation.

A World of Acronyms

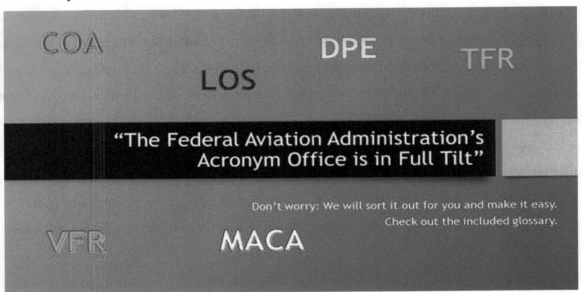

You would think the FAA has an entire office dedicated to developing acronyms to amaze and confuse us. Sadly, I think the volumes of acronyms has scared a lot of good pilots away from commercial sUAS flying. Considering this fact, we don't want you to worry, we will slowly introduce you to this new language. There is also a glossary included for your convenience.

Chapter 2

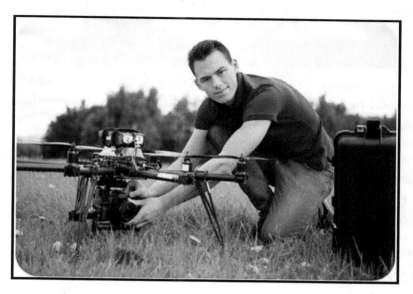

Your Remote Pilot Drone (sUAS) Certification

Remote Pilot (RP) certification consists of pre-qualification requirements followed by actions for earning Part 107 certification. The FAA has put together the legal framework in which commercial sUAS can operate in the sky above the United States, called the National Airspace System (NAS). This legal framework resides in the <u>Code of Federal Regulations</u>, specifically 14 CFR Part 107. The difficulty for people in the aviation is making sense of the "legal-ese" language used in the CFRs, or "regulations" or "regs" for short. Since the implementation of the Sport Pilot certificate an good effort has been made to make regulations more readable. Tip: When reading regulations, be sure to; follow all references to another regulation, watch for the words; "must," "should," "required," "mandatory," and the word "and or all."

The regulations contain information on not only certified, but also on operational requirements. To be able to apply (qualify) for RP certification (called Applicability and Eligibility) a person has to:

* Be at least16 years of age.

Note: Interesting you can be 14 year old to take the Part 107 Remote Pilot (RP) Knowledge Test (KT). I suppose the reason is that a 14 year old that takes the test on his/her birthday can submit it for an RP certificate on their 16th birthday (personally I wouldn't cut it that close). The test results are good for 2 years. Memory aid: 13 to register a sUAS, 14 to take the RP KT and 16 to fly commercially.

* Be able to read, speak, write, and understand the English language (English is the International Standard for aviation). Waivers are possible for those that have medical reasons for not complying with this rule.

* Be in a physical and mental condition that ensures safe flight, sometimes called "self-certifying" medical soundness. In the FAA words, "A person cannot operate a sUAS if they know or has reason to know that he or she has a physical or mental condition that would interfere with the safe operation of the small unmanned aircraft system" (CFR 107.17).

* No convictions involving alcohol or drugs. If there is a conviction, it must be reported to the FAA and may result in a minimum of a one year denial of an application (from the date of the act). The RP certificate holder may have their certificate suspended or revoked indefinitely. Refusal of submitting to a blood-alcohol test by a law enforcement officer is also grounds for suspension or revocation.

If a person meets all of these criteria, then he/she can take the Remote Pilot Knowledge Test (RP KT) at an FAA authorized testing center. The only disclaimer is that TSA will "vett" you prior to issuing your certificate. You will have 2 hours to answer 60 questions from randomly selected questions from the 300 plus question databank.

After passing the test (you will pass using this study guide) take the test results to an FAA Designated Examiner (DE), an Airmen Certification Representative (ACR) for a pilot school, a Certificated Flight Instructor, or other person authorized by the Administrator (which means FAA). Most people will use the services of a CFI for this task that will initiate the filling out an IACRA (FAA online application) form. A temporary Remote Pilot (RP) certificate will be issued (that expires in 120 days) until a permanent is sent via USPS from the FAA Airmen Certification Branch in Oklahoma City, OK. A TSA vetting process is also part of this process.

TC's Testing Tips: The most common mistake is not having a reasonable deadline for taking the test. A lot of folks will say to themselves, "I just need a little more time to study," and they never take the test or lose so much information that they do worse on the test. Unless you are able to lock yourself in a room in Bum Freaking Egypt it will take the normal person 1-2 months concentrated studying till 2-3 hours a day. You owe it to yourself to cut off all electronic devices, email notifications and social media. Multitaskers should note that studies prove concentrated study is more effective and efficient than multi-tasking. However, I do listen to "focus music," (it's a specific type of music, not listening to the Flat Duo Jets because you like their music). As your test day deadline approaches bump up your study time a week or two prior, which means to keep a check on your social and commitment calendar. Have all of your proper identification, "on your person" a week or two before the test. Get to the point of being able to answer the questions before you see the answer with 90% accuracy. Get plenty of sleep in general, but set a priority for the week off to be well rested, even if it means taking a couple days off work if necessary. The idea is to go into the test facility with your head full of information that is dying to get out. Stress level: I like to shoot for being a little nervous (sounds strange but it works) and well rested. Bring a simple calculator and plotter (more on this later). Come to grips ahead of time that there may be a few questions will not remember the answer, that's OK. Two hours is a lot of time to sort out hard questions. Take your time reading the questions, the FAA can be sneaky with questions, and answers, by changing them ever so slightly making them incorrect. Primarily stick with your gut reaction, too many people start second guessing their first answer and flub the test. If you are totally clueless on an answer, mark it for later. Interestingly, some of the preceding questions can give you the answer to the difficult questions and you have time for your mind to digest the question in your subconscious. You will come back to the question finding the correct answer. Worst case, use the process of elimination, usually there is a throw-away answer, as they call it, which means you now have a 50/50 chance of answering it correctly (always make an attempt). Check your stress level, breathe, take your time, and don't worry if people are completing the test before you. Remember, there are no points for finishing early. Now, take a deep break again, submit your answers and go get your passing test score. Positive attitude!

Identification of the RP Knowledge Test (KT)

When you take the Remote Pilot Knowledge Test make sure you have proper identification, less you be turned away. Proper identification contains the applicant's (this means you):

1) Photograph;

2) Signature;

3) Date of birth, which shows the applicant meets or will meet the age requirements of this part for the certificate and rating sought before the expiration date of the airman knowledge test report; and

4) Permanent mailing address. If the applicant's permanent mailing address is a post office box number, then the applicant must also provide a current residential address.

A list of acceptable documents used to provide proper identification can be found in Advisory Circular (AC) 61-65, Certification: Pilots and Flight and Ground Instructors (as amended). **Achieving a score of 70% (which I know you will do).** Tip: Always use data, like Advisory Circulars, directly from the FAA website to ensure you have the most current information.

Retesting after Failure

We are not going to spend too much time on this because you are going to pass with flying colors. In fact, we staked our name on it by providing a full refund of the cost of the study guide (just send us a copy of the test results). Like we said you are going to pass and you should encourage those not preparing adequately to realize they can't take the test for 14 calendar days after failure (not to mention another $150 to re-take the test).

Change of Name and/or Address

It is vitally important to keep up with notifying the FAA if you move to another location, have a name change or change your name. After 9/11 the DHS/TSA was hounding the FAA for the lack of good record keeping with respect to current Airmen data and aircraft registration information (covered in detail in Part 107.77). As they say, "stuff rolls downhill," so if the FAA has been in trouble regarding record keeping, you can bet it will roll downhill to Airmen and aircraft owners.

Keeping An RP Current

A pilot certificate never expires unless suspended, surrendered (turned in), or revoked. However, the use of the privileges of the certificate may prevent a pilot from using his/her pilot certificate. A manned aircraft pilot is required to do a certain amount of flying and a bi-annual Flight Review to keep his/her certificate current. RPs are required to take and pass a recurrent training course (or RPs can re-take the initial course) within the previous 24 calendar months. A calendar month is the end of the month in which the certificate (or last recurrent course) was passed. For example, you pass the RP KT (and are issued a temporary certificate the same day) August 29, 2016 (first day for the RP KT), you would need complete a recurrent test by August 31, 2018.

It's important to keep up with your recurrent training to keep you in the business of commercial RP flying. As a member of the RemotePilotAssociation.com, will will help you keep your RP current with recurrent training tools.

Aircraft Certification and Airmen Certification are two separate things.

Airmen certification and the standards thereof (Application and Eligibility) is different and separate than aircraft registration. This can be a source of confusion, for example, a person must be 16 years old to be a commercial Remote Pilot but at least 13 years old to register a sUAS. The difference is that a hobbyist can register his/her sUAS at 13 years of age. Let us say a 13-year-old waits until turning 16 years old and then wants to make money using his/her sUAS to make money. One 16 years old, he/she can take the Remote Pilot Knowledge Test, pass it, then register his/her sUAS as a commercial drone.

Your Aircraft Registration (CFR 107.13)

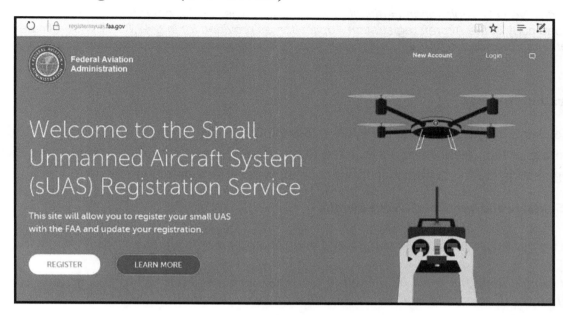

All sUAS must be registered with the FAA be https://registermyuas.faa.gov/ if they weigh between .55 lbs. and 55 lbs. be it for hobbyist or commercial use (including all accessories attached). A sUAS must be registered to a person that is 13 years of age or older. If the person is younger than 13 years old they can have someone 13 years old or older register the aircraft (like a parent).

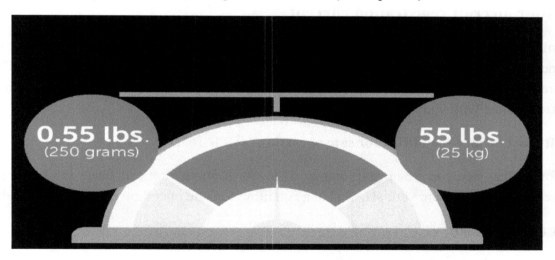

You must register your sUAS if it weighs between 0.55 lbs. and 55 lbs. (Source: FAA)

CFR 107.13 Registration is the regulation covering the requirement for sUAS aircraft registration by the aircraft owner. A copy of the registration must be retained within reasonable access (on the mission site) for inspection by the FAA. Part 107.13 refers to Part 91.203(a) (2) stating:

(a) Except as provided in § 91.715, no person may operate a civil aircraft unless it has within it the following:

(1) An appropriate and current airworthiness…

(2) An effective U.S. registration certificate issued to its owner or, for operation within the United States, the second copy of the Aircraft registration Application as provided for in § 47.31(c), a Certificate of Aircraft registration as provided in part 48, or a registration certification issued under the laws of a foreign country.

Registration Number

Upon registering your sUAS via the FAA website link at, https://registermyuas.faa.gov/ you will be issued a registration number that is required to be affixed to your aircraft. According to CFR 48.205 your registration number, unique identifier must meet the following requirements:

§48. 205 Display and location of unique identifier.

(a) The unique identifier must be maintained in a condition that is legible.

(b) The unique identifier must be affixed to the small unmanned aircraft by any means necessary to ensure that it will remain affixed for the duration of each operation.

(c) The unique identifier must be readily accessible and visible upon inspection of the small unmanned aircraft. A unique identifier enclosed in a compartment is readily accessible if it can be accessed without the use of any tool.

Keep your aircraft registration current

Like your Remote Pilot certificate, it is important to keep your aircraft registration current. As mentioned previously Airmen certification and aircraft registration and two separate issues. Changing the address associated with your RP (Airmen) certificate doesn't automatically change the registered address associated with your sUAS.

Aircraft must be in a "condition for safe operation" (CFR 107.15)

Don't forget that once you are up and running with your Remote Pilot certificate and have registered your aircraft that you ensure your sUAS is in a condition for safe operation prior to every flight.

Waiver (Part 107.200)

The existing regulation might not meet the needs of some operators. A Remote Pilot can apply for a waiver to, for example, fly at night, operate from a moving vehicle, fly over people and more. Below are the requirements for a waiver and allowable items the FAA will consider for waivers (however, you can request an exemption from any regulation if you have plenty of time and patience plus prove an alternative method of compliance).

§ 107.200 Waiver policy and requirements.

(a) The Administrator may issue a certificate of waiver authorizing a deviation from any regulation specified in § 107.205 of this subpart if the Administrator finds that a proposed small UAS operation can safely be conducted under the terms of that certificate of waiver.

(b) A request for a certificate of waiver must contain a complete description of the proposed operation and justification that establishes that the operation can safely be conducted under the terms of a certificate of waiver.

(c) The Administrator may prescribe additional limitations that the Administrator considers necessary.

(d) A person who receives a certificate of waiver issued under this section:

(1) May deviate from the regulations of this part to the extent specified in the certificate of waiver; and

(2) Must comply with any conditions or limitations that are specified in the certificate of waiver.

§ 107.205 List of regulations subject to waiver.

A certificate of waiver issued pursuant to § 107.200 of this subpart may authorize a deviation from the following regulations of this part:

Sec.621 107.25 – Operation from a moving vehicle or aircraft. However, no waiver of this provision will be issued to allow the carriage of property of another by aircraft for compensation or hire.

107.29 – Daylight operation.

107.31 – Visual line of sight aircraft operation. However, no waiver of this provision will be issued to allow the carriage of property of another by aircraft for compensation or hire.

107.33 – Visual observer.

107.35 – Operation of multiple small unmanned aircraft systems.

107.37(a) – Yielding the right of way.

107.39 – Operation over people.

107.41 – Operation in certain airspace.

107.51 – Operating limitations for small unmanned aircraft.

The process for applying for a waiver can be found on the FAA website at: https://www.faa.gov/uas/request_waiver/.

Certificate of (Waiver) Authorization (COA) for Governmental agencies and prior 333 Exemption Operators

RPs can apply for a waiver or a Certificate of (Waiver) Authorization (COA) to meet on-going unique flight operation needs. Here are quotes from the FAAs FAQ website page regarding COAs and 333 Exemptions:

I am part of a Federal/State/local government office – how can I fly a UAS to support a specific mission e.g. search and rescue?

You may either operate under the Part 107 rule, or you may be eligible to conduct public aircraft operations for which you would need to apply for a public Certificate of Waiver or Authorization (COA) for certain operations. Governmental agencies, like law enforcement, can take the Part 107 Remote Pilot certification test to fly under the general guidelines of Part 107.

Can my blanket Section 333 Certificate of Waiver or Authorization (COA) transfer to my UAS operating under part 107?

No. If you fly following the requirements of Part 107, you must comply with the operating provisions specified in part 107. Part 107 limits your altitude to 400 feet unless your unmanned aircraft is flying within 400 feet of a structure (in which case you may not fly higher than 400 feet above the top of that structure). Part 107 also limits your operation to Class G airspace unless you obtain FAA permission prior to the operation to fly in controlled airspace. The blanket COA issued with your Section 333 exemption is only valid if you continue flying using the conditions and limitations in your exemption.

Am I better off flying under the Part 107 rule or my Section 333 exemption?

It depends on what you want to do. UAS operators need to compare the conditions and limitations in their individual Section 333 exemption to the operating requirements in the Part 107 rule to determine which operating rules best address their needs.

Will I still need a COA to fly under the Part 107 rule?

If you already have a Certificate of Waiver or Authorization (COA), you can continue to fly under those COA requirements until it expires. Section 333 exemption holders may operate under the terms of their exemptions and COAs until they expire. Public aircraft operators such as law enforcement agencies, state or local governments, or public universities may continue to operate under the terms of their COAs.

If you don't already have a Section 333 exemption and associated COA, and you are not conducting a public aircraft operation, you probably don't need one now that Part 107 is out. Civil UAS operations flown under the new rules do not require the UAS operator to get a COA.

Need help? We would be happy to consult with you on applying for a 333 Exemption and COA. Contact me, TC Freeman at; mailto:tc@tcfreeman.com?subject=333 Exemption and COA.

Chapter 3

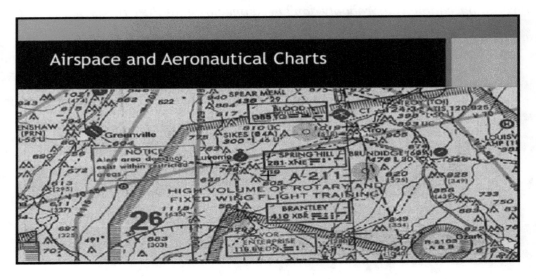

Airspace and You

Airspace is one of the most important chapters of this book and after reading this section you will never look at the sky in the same way. Just like the ground based system of roads and Interstates the air is filled with highways in the sky, no fly zones and busy corridors to major airports. We will help you see this new world, but airspace is one area that requires continued learning.

The concept of creating regulations pertaining to airspace is to keep aircraft from having midair collisions, be it in busy airspace associated with a large city or in military training areas. Keeping aircraft separated is not only the job of the air traffic controller, but also the pilot in making sure they have the in-flight visibility to stay safe and see other aircraft. A common misunderstanding with the general public is that aircraft are always under Air Traffic Control (ATC) supervision. There are many times a pilot will not be talking to an air traffic controller (flying in clear weather days, called Visual Flight Rules-VFR) and has to abide by minimum visibility standards. There are also areas of national defense and security that are off limits to airborne traffic. This is accomplished by establishing "no fly" or "limited fly" zones that are notated on a pilots aeronautical chart or given in special reports, called Notices to Airmen (NOTAMs).

The good news is that since sUAS only fly up to 400' above the ground (or aerial platform) there are a lot of areas in which to fly and fly safely. Manned aircraft are required to fly, at a minimum, 500' Above Ground Level (AGL) unless they are taking off or maneuvering for landing. As a sidebar, my major concerns with respect to having midair collision is with medivac helicopters, amphibious and agricultural aircraft (affectionately called crop dusters). Vigilance on the part of all aviators will ensure the continued health and safety of the industry.

The FAA has developed the airspace system that contains **Four Types** of broad categories **of airspace**:

***Controlled**, as in controlled by air traffic controlled, sometimes requiring pilot interface but not always.

***Uncontrolled**, but not without regulation. Occasionally ATC can assist pilots in this area, but not typical.

***Special Use Airspace** (SUA) consisting of military areas, sensitive to national security, etc.

***Other Airspace**: Local airport advisory (LAA), Military Training Routes (MTR), Temporary Flight Restriction (TFR), parachute jump aircraft operations, published Visual Flight Rules (VFR) routes, Terminal Radar Service Areas (TRSA), National Security Area (NSA), Air Defense Identification Zones (ADIZ), Flight Restricted Zones (FRZ) in the vicinity of the Capitol and the White House.

Don't worry, we will cover each of these, especially "Other Airspace" in this chapter. All Four Types of Airspace are under the larger umbrella **Two Categories of Airspace; Regulatory and Non-Regulatory**. This is a little confusing since all airspace is regulated in some form even if uncontrolled, but this is what the FAA wants you to know.

To begin developing your new 3D vision of the sky, I refer to the Airspace Classification Chart below is a good way to see Airspace.

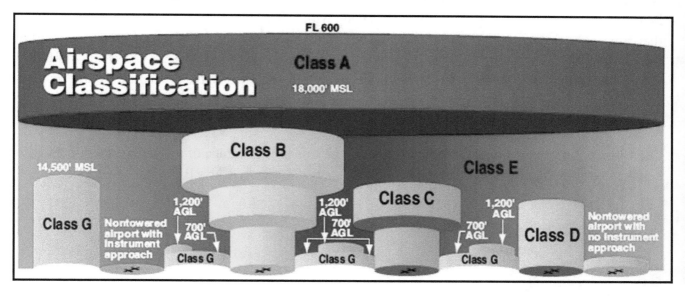

Graphic A (Source: FAA)

Study the Airspace Classification chart a little closer and it will become obvious that a lot of the airspace in the National Airspace System (NAS), as the FAA likes to call it, doesn't pertain to the average sUAS pilot. However, airspace is probably one of the weakest areas of knowledge for the manned pilot and are the reason for many violations. Therefore, it is extremely important to become very knowledgeable about the airspace in which you fly. Let us jump right into learning about it.

Class A

Class A Airspace is for flight operations above 18,000' MSL, the land of airliners, corporate aircraft and personal turboprop aircraft. Memory aid: A Airspace is for the Airliners way up high.

Note: the next four Airspace Classifications are associated with airports that have Air Traffic Control Towers (ATCT). The FAA states that request to fly in these areas (B, C and D Airports) must be made with the FAA via an online form available at: https://www.faa.gov/uas/request_waiver/. Not to be the bearer of bad tidings, but this process can take up to 90 days according to the FAA.

Class B

Class B Airspace is synonymous with big cities and the very large airports they serve. Class B Airspace and the airport they are associated are the busiest in the country like; Los Angeles (LAX), Atlanta (ATL) and Chicago (ORD), just to name a few. Due to the massive amount of flight operations Class B Airspace extends out 30 NM from the center of the airport and rises vertically to 10,000 AGL (in a generic, non-customized, design). The entire radius consist of various layers that can best be seen via the Airspace Classification graphic. For visualization purposes some refer the construction of Class B Airspace (and C) as an upside-down wedding cake. The good news for UA operators is that due to the laying of airspace there is a lot of room to fly, even within five to ten miles of the airport. Having said that, always be on the lookout for aircraft traffic in this busy area and be aware of numerous satellite airports that surround the main Class B airport. The FAA has a generic design (30 NM and 10,000' AGL) but most Class B Airspace is customized for the specific needs of the area (such as local geography and satellite airports). Prior permission from the Class B, Airspace Air Traffic Control Tower (ATCT) is mandatory.

The Class B Airspace on the chart (below) doesn't give us the advantage of a 3D view, therefore a pilot needs to be able to know what the various layers the wedding cake Class B consist of altitude-wise.

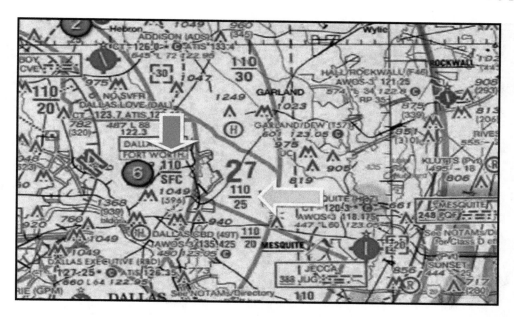

Figure 25, Class B altitude coverage in various segments

Figure 25 is a portion of the Dallas-Fort Worth (KDFW) Airport Class B with its layers denoted by light blue lines, the KDFW Airport Northwest of Area 6. Each segment of blue lines is a different layer of the airspace. Notice to the East of Area 6 has the numbers 110/SFC (blue arrow), which means the Class B extends from the Surface (SFC) to 11,000' MSL. Further to the Southeast are 110/25 (orange arrow), meaning the Class B extends from 2,500' MSL to 11,000' MSL. Practice looking for other layers and the altitudes they cover. The obvious key to this exercise is to carefully follow the blue lines and find its corresponding altitude information. We encourage you to use the expanded view chart, called a Terminal Area Chart (terminal area, that sounds lovely) that is available for every Class B Airspace in the US.

Class C Airspace

If you think Class C Airspace looks a lot like Class B Airspace minus some of the layers, you would be right. Class C airports are not as busy and are typically associated with medium size cities. They are typically 10 NM in diameter and are 4000' AGL (when constructed in the generic fashion). Memory Aid: The C in Class C Airspace stands Cities. Graphic A can give you a good idea of Class C in 3D, but let us look at Class C on an aeronautical chart (Figure 23, below).

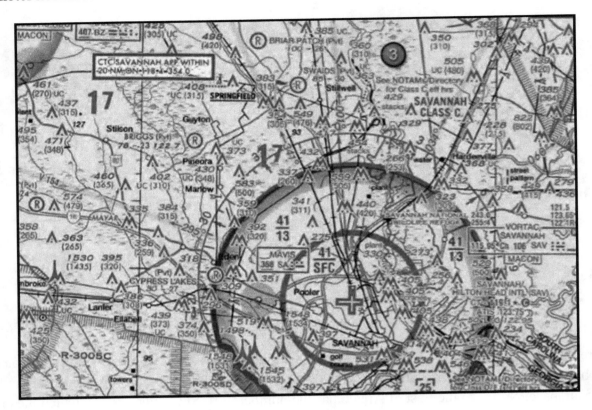

Figure 23. Not to scale.

The Savannah /Hilton Head Intl Airport (KSAV) is a Class C Airport and as you can see is a lot easier on the eyes with only two layers. The inner layer is from the Surface (SFC) to 4,100' MSL (see the 41/SFC Northwest of the airport, inner circle). The outer ring (see the number 41/13 further to the Northwest inside of the outer ring) is from 1,300' MSL to 4,100' MSL. Besides the double layer of airspace Class C can be identified by it's magenta color rings. Basically, it is a standard constructed Class C, but the SAV Class C has a notch in the Southwest corner of the outer ring to make space for military airspace denoted by the (Restricted) R-3006D inside of blue hatched lines (more on, Restricted Airspace later). KSAV is a control towered airport indicative of Class C Airports, you can also verify this by the airport information box on the East side of the airport under the airport name see Control Tower (CT) 119.1.

Since Class C Airports have an Air Traffic Control Tower (ATCT), like the example of Savannah, you will need prior permission from ATC to fly in this airspace.

Class D Airspace

Class D Airspace is at the end of the chain when it comes to the classification of airspace that have an ATCT. The least busy of air traffic controlled airports with only one layer of airspace to be concerned, a cylinder like area that, in its generic form, extends out 5 NM from the airport and stands 2,500' AGL vertically. Fortunately, Class D makes the job easy to pick out the dimensions of the airspace (see graphic below).

26

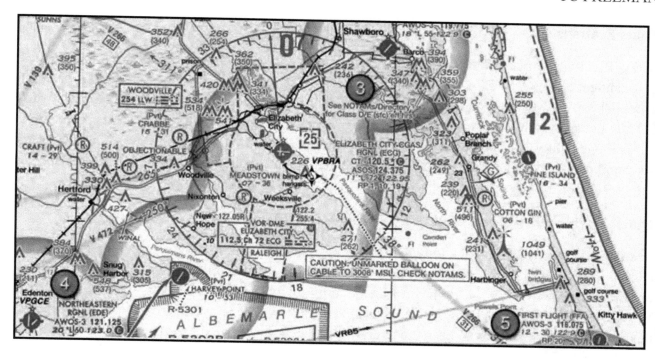

Figure 20, Area 3

Notice the blue dashed lines that make up the Elizabeth City/US Coast Guard (CG) Regional Airport (KECG) Soutwest of Area 3. Northeast of the KECG Airport inside of the blue dashed lines is the number 25 inside of a blue dashed line box, which means the Class D Airspace extends from the Surface (SFC) to 2,500' MSL. Even though Class D Airports are less busy than B and C airports it has an ATCT, note the Control Tower (CT) 120.5 below the airport name. Class D Airspace requires prior permission to fly in this airspace.

Next we are going to talk about E and G airspace, which are in the same alphabet family of airspace but do not contain airports with an ATCT. You might be asking, "where is the F Airspace?" Well, there is no F'in airspace…and that is all I will say because this is a G rated publication. Anyway, back to E and G. If we have A Airspace for Airliners, B Airspace for Big city airports, C Airspace for good regular Cities, and D Airspace bringing up the rear representing the smallest ATCT airports, what about all the airspace in-between airports? This is where Class E and G Airspace comes in. But first.

Gaining ATC Permission to Fly in Class B, C and D Airspace

The FAA has said that prior permission must be given by the ATCT of the Class of ATC airspace you want to fly in. The process to do this will not be with the specific ATCT but through the FAA website at: https://www.faa.gov/uas/request_waiver/. The first block of the online form is under "waiver" is the selection for "Operation in certain airspace" from Part 107.41. Approvals can take up to 90 days for approval. Don't kill the messenger.

Class E Airspace

The difficult challenge with describing Class E Airspace is that it isn't obvious when looking at the aeronautical chart. Class B, C and D Airspace are dead giveaways and can be pointed out easily on a chart. Class E is what you might call in-between, or filler, airspace that is in-between airports with an ATCT. Looking at the Airspace Classification (Graphic A) will give you the best 3D representation of Class E. Generally, this airspace may start at the ground, 700' AGL or 1,200' AGL depending on the surrounding airspace. This is a good time to talk about manned aircraft pilots flying under Instrument Flight Rules (IFR).

Quick Sidebar: Instrument Flight Rules (IFR) Flying

Instrument Flight Rules (IFR) Flying is done by manned aircraft primarily during poor weather conditions. To cope with not being able to see well in limited visibility conditions, in the clouds for example, a pilot will use the flight instruments only to fly the aircraft. Clear weather pilots, called Visual Flight Rules (VFR) flying, the pilot can clearly see outside to use external visual references like the horizon, ground and sky to keep the aircraft flying upright. When these visual references are taken away the pilot has to revert to using the aircraft instruments to keep him or her flying. Since pilots flying IFR cannot see outside, they are under ATC assistance for 90% of the flight. The remaining 10% is for an instrument approach into an airport, and this is where it gets interesting. Instrument pilots on approach to an airport are sharing airspace with VFR pilots unless the airport is really "socked in" with very low visibility (clouds, fog, etc.) and VFR pilots can't fly. This is the case the designers of the airspace, said we need stricter weather minimums, for example 1 mile visibility and staying clear of the clouds, to safeguard against a midair collision. Sharing airspace and protecting pilots from midair collisions is an important point when it comes to understanding Class E and G Airspace.

Now Back to Class E Airspace

One of the easiest forms of Class E Airspace to see on an aeronautical chart is E that extends to the ground at airports with an instrument approach (see below).

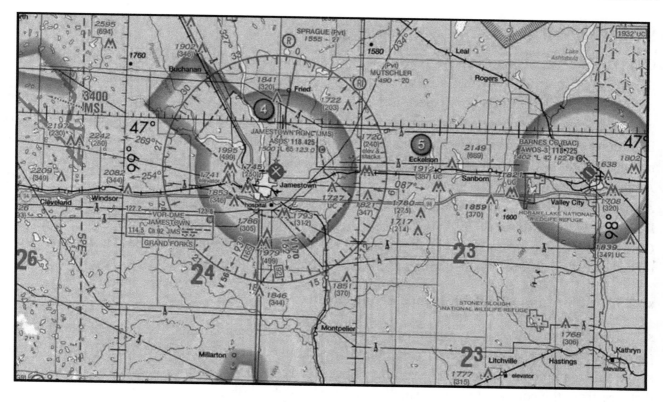

Figure 26, Area 4

The Jamestown Regional Airport (KJMS) is a great example of Class E Airspace, noticed the magenta dashed lines, meaning that Class E starts at the Surface (SFC). Also notice that this airport doesn't have an ATCT but does have instrument approaches. How can you tell? Check out the key hole extensions of the magenta dashed lines that are roughly aligned with the runways. The FAA is using more strict visibility standards to protect aircraft on approaches from VFR aircraft and other IFR aircraft. "How far does it go up vertically," you might ask? In the absence of a number (listed on the chart) Class E goes all the way up to the base of Class A (18,000' MSL or Flight Level 180 more correctly).

Looking again at KJMS Class E you might wonder what happens when you go outside of the dashed line to the magenta shading? As we said previously, inside of the magenta dashed line Class E starts at the ground, just outside of the line to the magenta shaded area Class E goes up to 700' AGL in an effort to still protect those aircraft on approach. Outside of the magenta shaded area Class E goes to 1,200' AGL. Class E will continue in-between airports or other airspace of significance (see Figure A for a visual image). The airspace below 700' AGL and 1,200' AGL is open for the category of airspace called G-Airspace.

Note: Don't fall into the trap of mistaking a Class D Airport with its blue dashed lines and ATCT for Class E with magenta dashed lines. Class E Airports do not have an ATCT but have instrument approaches. These approaches are coordinated by remote ATC facilities like a close-by large airport or an ATC Center facility.

Class G Airspace

Class E, and especially G Airspace, is where UAS thrives. Class E is considered "Controlled Airspace," think "ATC Controlled," with IFR services that are offered as an option to manned aircraft pilots but not off-limits to UA. Class G Airspace is called "Uncontrolled Airspace," perhaps due to the limited amount of coverage ATC has over this area, but they still coordinate IFR approaches (and VFR aircraft) into this area. It's important to know that Uncontrolled does mean without rules (regulations). This is not my definition of uncontrolled but I digress. Use Graphic A to get a visual image of Class G Airspace, now take a look at Figure 26. The areas in-between the magenta shaded circles is Class G Airpace that starts at the ground and goes up to 1,200' AGL. Transitioning inside of the magenta shaded line Class G drops down to 700' AGL unless there is a magenta dashed line where Class G stops and Class E goes to the ground. This is all in an effort to keep aircraft from running into each other. Next up is an entirely different type of airspace for the military called, Special Use Airspace.

sUAS Visibility and Cloud Clearance Requirements

sUAS aircraft have their own visibility and cloud clearance requirements like manned aircraft. The good news is that it is just one set of requirements, unlike manned aircraft. CFR 107.51 states that sUAS are required to have 3 SM visibility. Specifically the FAA defines it as the, "average slant distance from the control station at which prominent unlighted objects may be seen and identified by day and prominent light objects may be seen at night (waiver).

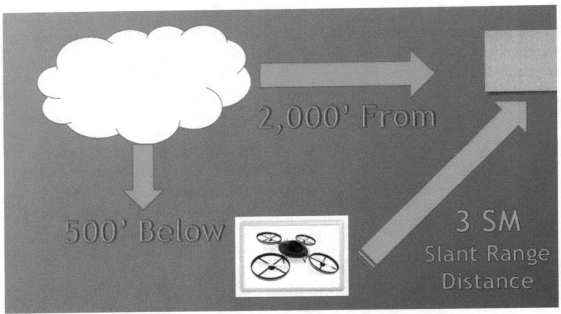

Visibility and Cloud Clearances required via CFR 107.51

30

Here is an example of visibility and cloud clearance requirements: You arrive on the flight mission site and there are clouds at 700' AGL. How high can you legally fly? Answer: 200' AGL (700' – 500' = 200' AGL).

Special Use Airspace (SUA)

What makes Special Use Airspace so special? It is that it has restricted to limited use of pilots, I'm not sure that sounds so special. Never-the-less, the major function of SUA is a good one, to allow the military to have training exercises and to protect sensitive areas of national security. Let us take them one by one starting with Military Operations Areas (MOAs).

Military Operations Areas (MOAs)

Military Operations Areas (MOAs) are areas in which the military holds training exercises, like aerial dog fighting and simulated ground attacks. When the military is holding exercises they issue a Notice to Airmen (NOTAM), available through the weather briefing service at the Automated Flight Service Station (AFSS), that the specific MOA is "hot." After the exercises are over the MOA is said to be "cold," or inactive. Details about the specific MOA can be found around the edge of the aeronautical chart that includes; altitudes, times of operation, and the controlling agency contact. What is good about MOAs is that you can fly through them at any time, "hot" or "cold." The FAA says to exercise extreme caution but they are fair game. Is it wise to fly through a "hot" MOA? Probably not, but heck you can be part of the fun (just kidding). Personally, I would wait until the area is "cold." See the MOA below inside of the magenta hatched lines below.

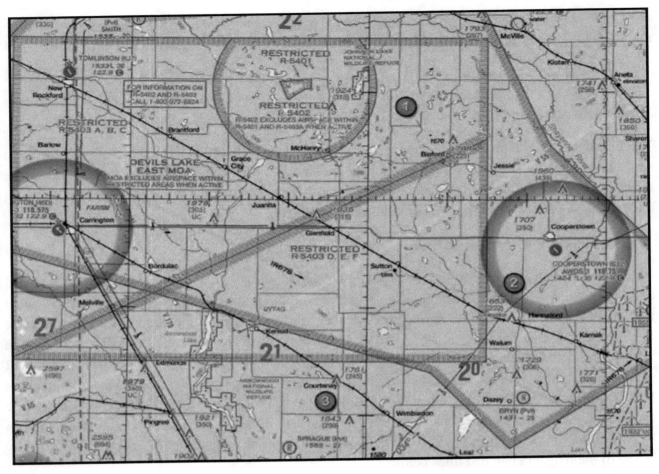

Figure 26. Not to scale.

The magenta hatched lines of the MOA look as confusing as the Dallas-Fort Worth (KDFW) Class B Airspace with the mix of magenta and the blue hatched lines in a Restricted Area (that is just layering), which is up next.

Restricted Areas

Restricted Areas on Figure 26 look very similar to the magenta hatched lines of an MOA but with blue hatched lines. On the edge of the chart you can find more information about the altitudes covered, times of operation and the controlling agency contact. Restricted areas contain much of the same military training exercises that are stepped up a notch with live firing. Needless to say, access to these areas is limited, however, it is worth contacting the controlling agency to ask for permission to fly in this area. If the area is "cold" access is typically granted to manned aircraft flying by IFR and to a select number of VFR pilots.

Alert Areas

Alert Areas are also areas of military training, specifically a high number of flight training activities or other unusual aeronautical activities. Pilots should exercise caution in these areas at all times. FAA guidance suggests during a weather briefing by Automated Flight Service Station (AFSS) you request information on activities in Alert Areas (Figure 69, Area 6).

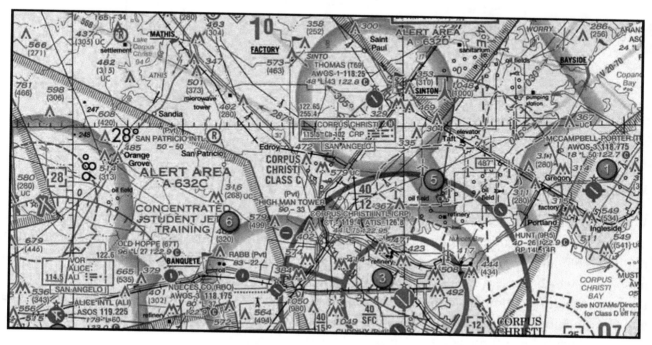

Figure 69, Area 6. Not to scale.

Notice that an MOA and Alert Areas both use magenta hatched lines to denote their respective airspace. Similarly, they operate under the "exercise extreme caution" provision. The difference is that Alert Areas are not as likely to have a NOTAM regarding it as being "hot." The lesson is to always be on your toes in Alert Areas and consult the edges of the chart for details about; times, altitudes and the controlling agency contact. Notice the statement below the Alert Area A-632 text stating "Concentrated Student Jet Training." Next are areas of national security called Prohibited Areas.

Prohibited Areas

As the name implies, Prohibited Areas are exactly that, prohibited. These areas contain properties associated with national safety and security, for example Camp, David (see below).

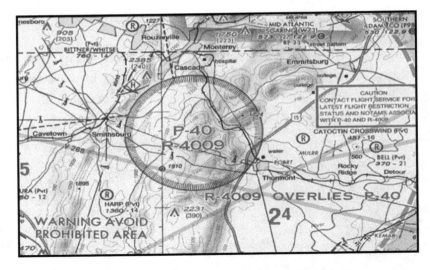

Prohibited Area (source: public domain)

Warning Areas

Warning Areas are typically found off the coast of the United States, starting 3 NM from the coastline and may be considered international waters, domestic or both. This means the US government might not have sole jurisdiction over the airspace. For flight operations it means this is like the wild west of airspace in which dangers can exist most anywhere. We realize that UA are used for ships and boats so the best recommendation is to exercise extreme caution, contact the controlling agency if applicable, and keep in mind that there is an extra level of risk involved with operating in a Warning Area (see the Warning Area below off the coast of Virginia (Figure 20, Area 7).

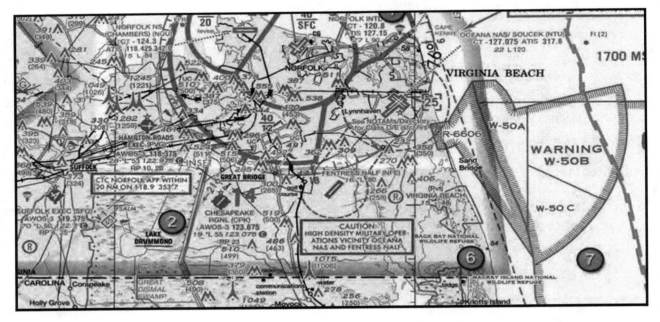

Figure 20, Area 7. Not to scale.

Controlled Firing Areas (CFAs)

It is interesting that Controlled Firing Areas (CFAs) are considered a type of SUA, but are not listed on the aeronautical chart. Never-the-less, they are typically integrated into another type of SUA, such as a Restricted Area, and are staffed with professional "spotters" that are tasked to suspend operations if aircraft are sighted.

Other Airspace

Other Airspace contains several types of airspace that are less seen, but we will touch on a few of these items. Other Airspace is; Local airport advisory (LAA), Military Training Routes (MTR), Temporary Flight Restriction (TFR), parachute jump aircraft operations, published Visual Flight Rules (VFR) routes, Terminal Radar Service Areas (TRSA), National Security Area (NSA), Air Defense Identification Zones (ADIZ), Flight Restricted Zones (FRZ) in the vicinity of the Capitol and the White House.

Military Training Routes (MTRs)

Military Training Routes (MTRs) are listed on the aeronautical chart as gray lines (Memory aid: think of gray camouflage military aircraft, see Figure 21, East of Area 2, IR644 – 649). These routes allow military aircraft to fly longer cross-country type training flights, either VFR or IFR, at speed in excess of 250 knots. Each gray line MTR begins with IR or VR followed by a series of numbers. MTRs with four digit numbers is flying at an altitude below 1,500' AGL. A IR or VR number that is three digits are routes flown at 1,500' AGL or higher.

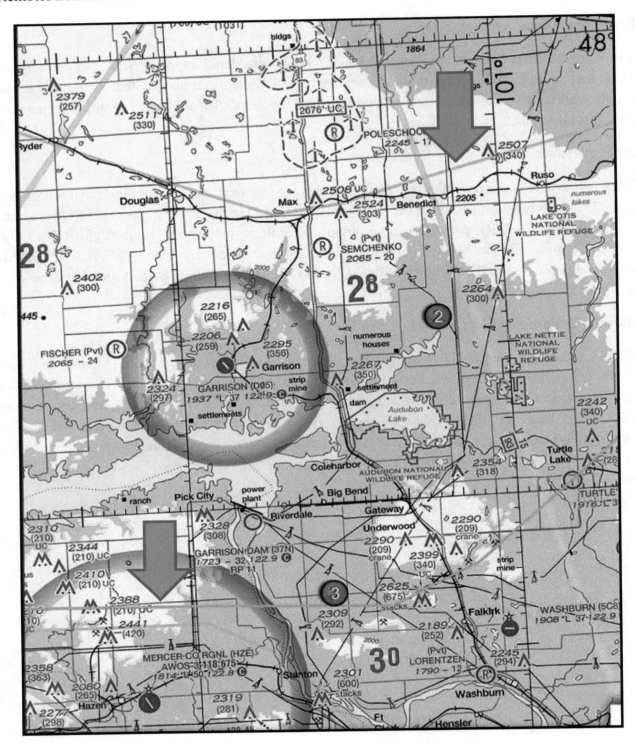

Figure 21, East of Area 2. Not to scale.

Temporary Flight Restrictions (TFRs)

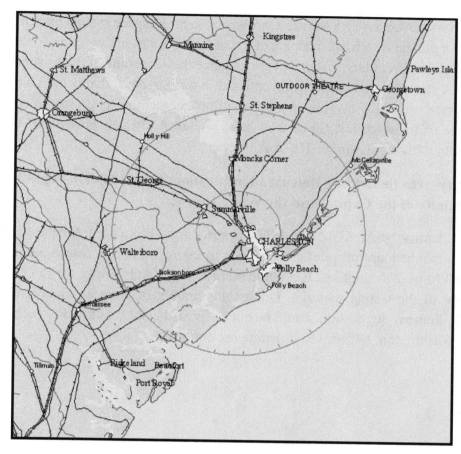

Temporary Flight Restriction (TFR) Map (source: public domain)

Temporary Flight Restrictions (TFRs) should have its own chapter in this book. TFRs were born in the aftermath of the post-9/11 terrorist attacks, but have been expanded from its origins of quickly protecting various airspace sensitive to national security and for the movement of the President (Vice-President and other public figures). TFRs now include protection of disaster areas, space operations and high profile events. All pilots are required to check TFRs prior to flight, which can be done via the AFSS weather briefing service or online. Personally, I like to go direct to the source by going to the FAA website for TFRs using their graphic map feature at: http://tfr.faa.gov/tfr_map_ims/html/ew/scale3/tile_4_2.html. The text version is very cumbersome and time consuming to me, but to each his or her own.

Note: One issue that has created problems for pilots is the movement of the President, Vice-President and other officials, especially during an election season. When a person that is covered under a TFR, such as the President, is traveling the TFR moves with him or her. Unfortunately, this has caused a spike in TFR violations and continues to be a challenge in dealing with the movement of officials and public figures. The best advice is to exercise due diligence by getting TFR information as close to the flight mission time as possible.

Terminal Radar Service Areas (TRSAs)

Terminal Radar Service Areas (TRSAs) are a throwback to the airspace system prior to the International Civil Aviation Organization (ICAO) alphabet classification system we have today. Despite their ominous bold black circles of large swaths of the airspace they are some of the least busy ATCT served airports. The FAA states that participation is optional in this airspace, that is until a pilot is within 5 – 10 NM of the airport area. However, most manned pilots contact the ATC prior to entering the TRSA to take advantage of their service. Since they have a control tower it is required UA operators get permission prior to a flight mission in TRSA Airspace.

National Security Area (NSA), Air Defense Identification Zones (ADIZ), Flight Restricted Zones (FRZ) in the vicinity of the Capitol and the White House.

I would closely identify NSA, ADIZ and FRZ Airspace with Prohibited Airspace. There are certain procedures that a manned aircraft pilot can do, such as taking an online course, before being allowed to fly within 60 NM of Washington, DC. While there is most likely a way for UA to gain access it is outside the scope of the testing material. Due to the amount of airspace violations I have seen by manned aircraft pilots my suggestion would be not to fly in the areas (and other such areas). Below is a picture of the Washington, DC area to illustrate my concern.

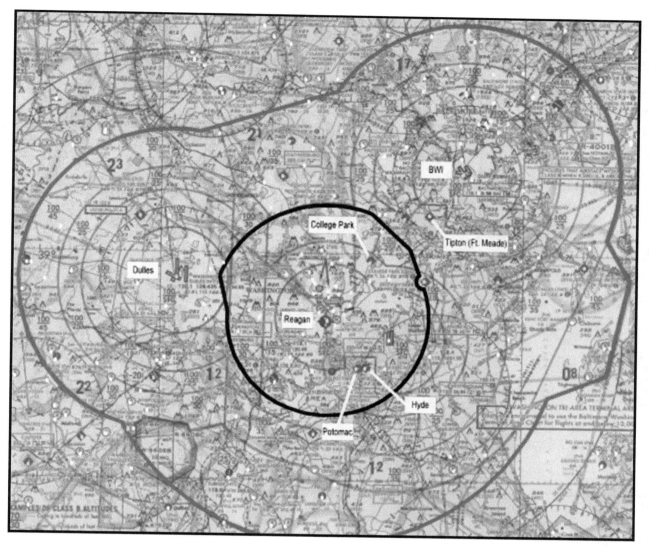

Washington, ADIZ (Source FAA article: "TFR Airspace Obstacles and TFR Trivia). Not to scale.

If you want to "geek out" on this important type of SUA read the following article by the FAA: http://www.faa.gov/pilots/safety/notams_tfr/media/tfrweb.pdf

*Aeronauical Chart Legend, Use It! It's packed with information and included with your testing materials.

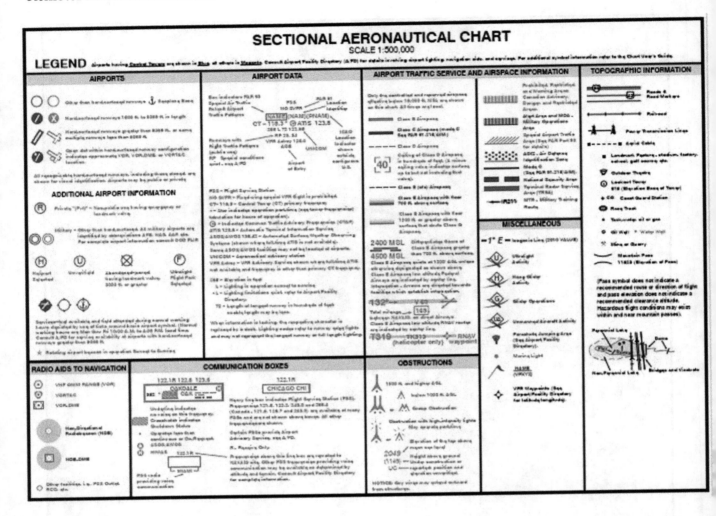

Legend 1

In the previous text we broke down the major parts of the National Airspace System (NAS) to ease you into reading a full chart. We highly encourage you to take advantage of the chart Legends that are provided with the aeronautical chart and included in the testing materials. To get a full breakdown on chart symbols go to: http://www.faa.gov/air_traffic/flight_info/aeronav/digital_products/aero_guide/#VFR_Symbols

We are going to cherry pick a couple of important items from charts:

Towers and Lighting

Towers (antenna) have the larger symbol (see below) for towers that are 1,000' Above Ground Level (AGL) and higher and small symbols for towers below 1,000' AGL. All towers above 200' AGL are required to be charted (so be careful out there, especially with the mass of cell phone towers going up) but some towers used high-intensity lights (while sometimes part-time or via proximity sensor). The main number in bold is the Mean Sea Level altitude, which works with manned aircraft altimeters (GPS altimeters can be close but don't bet the farm on it). The lower figure is the altitude stated as AGL. If the text says "UL," the elevation is unknown (or unlisted as a memory aid). Note the important "notice" at the bottom of the graphic about guy wires. This is one of the main hazards to aircraft as they can protrude from the main structure 2000'.

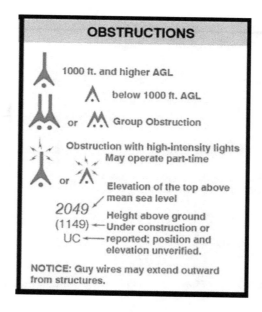

More on Group Antenna Obstructions

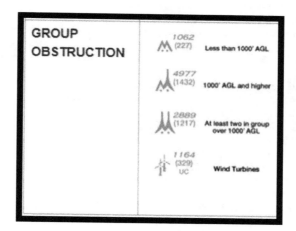

I can see how people can get confused with a double group of light, high towers, looks a lot like a starfighter from the *Galaga* video game (shows you how old I am, kicking it old school).

Airport Supplements

Airport Supplements, previously known as the Airport Facility Directory (AFD) contain information about the airport that won't fit on the aeronautical chart such as, airspace type, services offered, runways, distances from the associated city, etc. I would think a better name is an airport guide. However, the challenge is that this, like many other FAA materials, it is in code (shorthand). Fortunately, there is a decoder available with the Airport Supplements and your provided testing materials as a Legend. Here is an example of an airport supplement followed by a decoder excerpt.

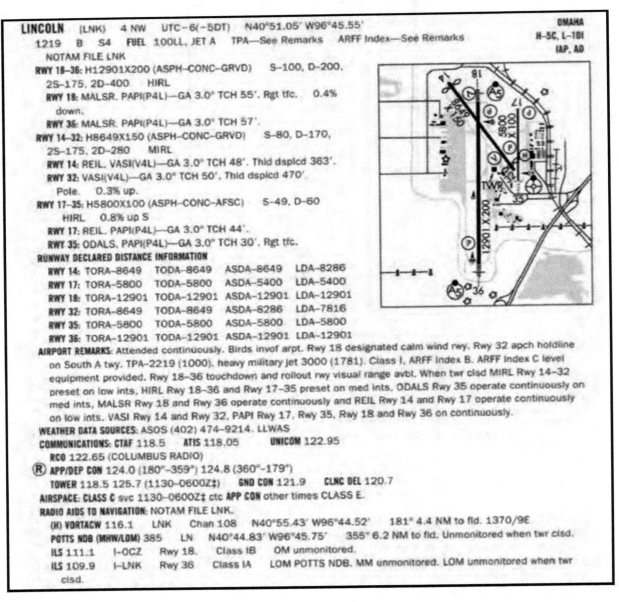

Figure 53

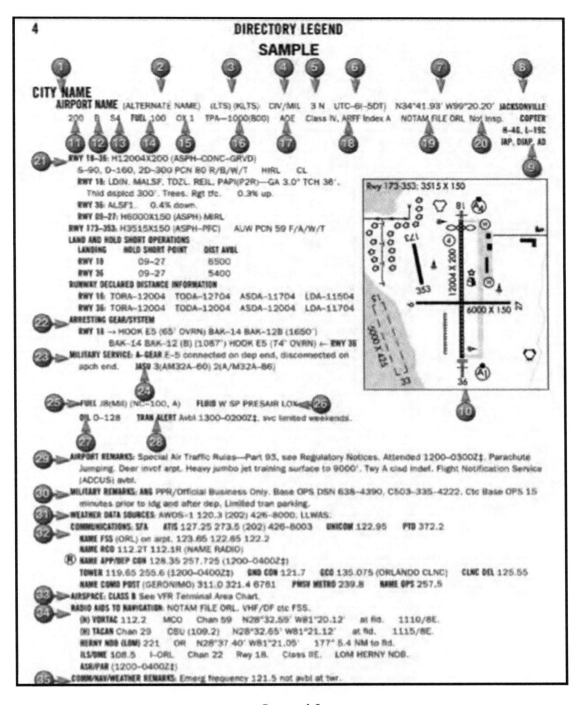

Legend 2

Don't worry, even the most seasoned pilots have trouble decoding all of the information on the Chart Supplements. Only a fraction of the information actually pertains to the UA pilot. The key is to use the de-coder in Legend 2 that has referenced numbers (in magenta) with a full explanation later in the document.

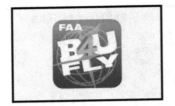

B4UFly App (source: FAA)

The FAA understands that learning charts and airspace takes a long time to master and created the B4UFLY App for Smartphones to help. Unfortunately, the reviews were not the best for the app, but it can be worthwhile to download (http://www.faa.gov/uas/where_to_fly/b4ufly/). Here is a screenshot:

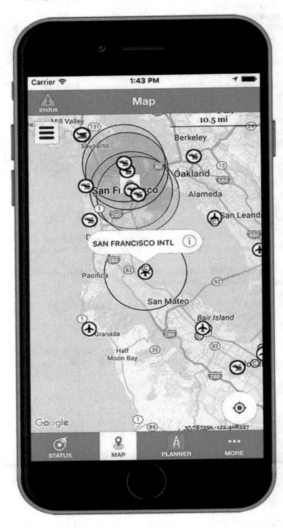

B4UFLY App screenshot (source: FAA)

One of the advantages of the app is that you can quickly see any airport and airspace issues at one glance. However, a complaint is that the screen is too cluttered because it includes warnings about all airports. When you look closely at an aeronautical chart you will see that in most States there are

44

hundreds of airports, from small grass airfields to ultra-busy Class B Airports that can create a barrage of information. I'm sure the app will become more efficient over time as more feedback is received by users.

Finding Latitude and Longitude

Finding Latitude and Longitude (lat/long) in the modern age of GPS technology is a tough sell but it is important, not only for the test but as an initial planning tool and backup. GPS signals can be blocked or inoperative, batteries quit working on cell phones and laptops, etc. You may have studied latitude and longitude in school, so let us reach back far in the data banks and pull that information out once again. We are going to focus on the lat/longs over the US since that is what the Remote Pilot test is primarily focused.

Basics

Lines of latitude run horizontally across the Earth and longitude lines are vertical (Memory aid: longitude runs a LONG, or longways, across the globe. Lines are used as a reference stated in degrees from a reference point (or origin, zero point). The zero-point latitude is the equator and for longitude is the Prime Meridian (Greenwich, England).

Major degree lines of latitude and longitude are separated by divisions of 60 minutes. The tricky thing in aviation is that there is an extra reference line in-between degrees that is not marked, this is the 30-minute mark between degrees. For example, say you are on pointing at the chart to the 70° line of longitude and move West (to a greater number) in-between 70° and 71° (there is a non-numbered longitude line here). The blank (or un-numbered) line stands for 70° and 30 minutes. A common mistake is to assume the next reference line is the next whole number, 71° for instance.

Lines of Latitude

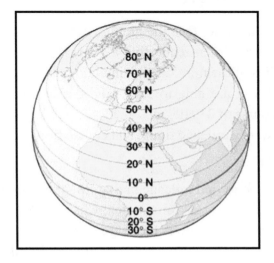

Lines of Latitude (source: web.gcc.edu)

Lines of latitude are lines numbered in degrees that run horizontally across the US and the Earth. The zero-degree point for latitude is the equator and numbers grow greater as you travel the closer to the North Pole (from a US perspective). In other words, latitude number increase in number when traveling from the South to North over the US.

Lines of Longitude

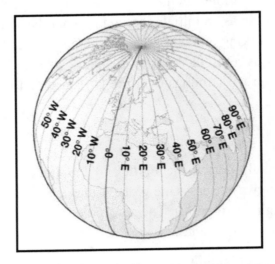

Lines of Longitude (source: web.gcc.edu)

Lines of longitude are lines numbered in degrees that run vertically across the US and the world. The zero-degree point is at the Prime Meridian (Greenwich, England) and increasing in degree number when traveling from East to West in the US.

Put them together and you have an X-Y axis grid system, YAY!

Determining Direction Using an Aeronautical Chart and Plotter

This is a great time to invest in a plotter (see the graphic below), a ruler like device used to calculate mileage and to determine courses. When requesting a waiver to fly in Class B, C or D Airspace the FAA will ask about the specifics of the mission site, including lat/long and perhaps with respect to the airport or navigational aid. If you are able to state (and depict on a chart) that the site is 5 miles Southwest of the airport on a course of 230 degrees, the FAA will know that you have your act together. More times than not they will want a mileage figure and course from a navigational facility, but figuring courses is a good start.

The ruler part of the plotter is used to find mileage in Statute Miles (SM) and the Nautical Miles (NM). Caution: Be sure you are not using World Aeronautical Chart (WAC) because the mileage scale will be incorrect for most plotters, plus the scale is too large for practical use. The curved part of the plot consists of one-half of compass numbers. Remember, there are 360 degrees in a circle with cardinal heading of 090 (zero nine, zero) for East, 180 degrees for South, 270 degrees is West, and 360 degrees (or 000°) is North (see the illustration below).

If a course is a totally foreign concept, I like to use Tony Hawk, the famous skateboarder, to describe the concept of course. Tony and his trick skater peers are always competing to see who can pull off the "sickest trick," (sick is good in this case) by doing 180's, 270's and 360's, as in degrees of turn. They actually do 720's but this won't help us in this case, back to the lesson. The skater can turn in one degree increments all the way around in a circle. Let us say the skater is rolling down the centerline of a road heading perfectly North, which is 0° or 360°. If does a quick total reverse, of course, heading back the exact opposite course he was headed, now he is traveling South or 180°. Say he now makes a 90 degree turn to the right, 180° + 90 degrees (as I degrees of turn) = 270°. Starting to get the picture?

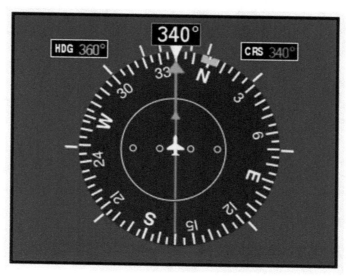

Understanding course: Remember there are 360° in a circle (Source: FAA)

Notice in the graphic below that the selected course is placed under the ruler part of the plot. Next, a line of longitude is found the reference point (red line) or the answer. The plotter is designed with multiple numbers on the compass scale to allow a pilot to figure direction using lines longitude or latitude depending on which way the course line is drawn.

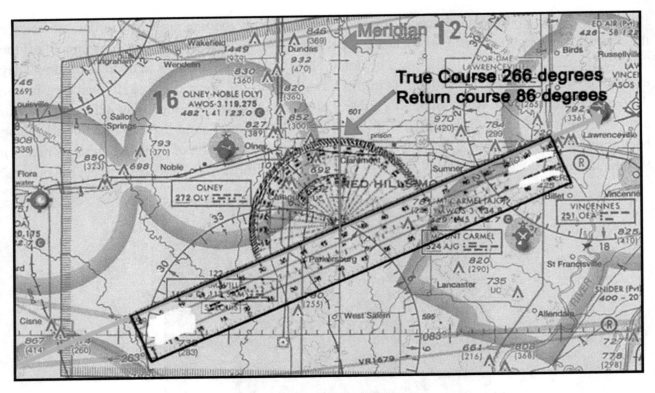

Using a Plotter to find True Course and Distance (source: siue.edu)

What About Direction from a VOR Navigation Facility?

The FAA may ask you for the location of your mission site with respect to a close by navigation facility called a Very *High Frequency* Omni Range (VOR). A VOR is a pre-GPS ground based navigational system with facilities spread out across the country to allow pilots to navigate directly to and from these physical sites. The easiest way to find VORs is to look for their associated compass rose (see the chart graphic below). There is a symbol for the VOR found on the aeronautical chart legend as another reference. However, if the VOR is located at an airport there will not be a VOR symbol, but the compass rose it still included.

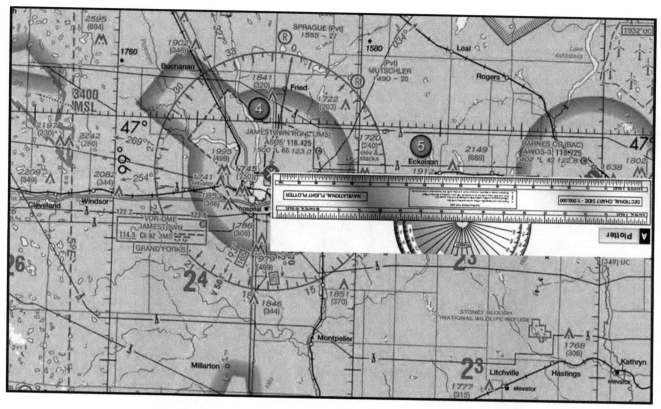

Figure 26 with super-imposed plotter to plot from a VOR. Not to scale.

We will use the chart and plotter below to find the course, called a radial, using the Jamestown VOR. Our mission site is over the small town of Eckelson and the Jamestown airport manager would like to know its radial and distance (find distance is going to be a guess due to not having a correct scale on the super-imposed plotter). Unlike calculating course using latitude and longitude the only thing you need to do is to lay out a line using the ruler side of the plotter, and draw out from the center of VOR (located on the airport in this case, see below) to the mission site. First, use the mileage side (ruler) to determine the mileage (I'm going to estimate 13 NM based on the Surface based Class E being 5 NM). Next, follow the line cross the compass rose for the VOR and determine the radial. Tip: The tick marks are 5 degrees apart, with the longest marks being 10 degrees. Additionally, there are reference degree numbers on parts of the compass rose, which in our case is 080° radial. Talking to the Jamestown Airport Manager you state, "The mission site is 13 Nautical Miles out on the 080-degree radial of the Jamestown VOR." Very professional communication.

Note: Using the compass rose on the VOR to find a course, called a radial that is actually Magnetic Course. Magnetic Course is calculated using True Course, which we spoke about in the initial plotter exercise course, and is corrected for Variation. What is Variation? Funny you should ask.

Magnetic Variation

Magnetic Variation is listed on the aeronautical chart as long magenta dashed lines that travel semi-vertically on the chart (see the graphic and chart below) that is used to illustrate the conversion factor of True North from Magnetic North. This calculation is called Magnetic Course and is used by manned aircraft pilots for cross-country flights. Before we show Variation on an aeronautical chart, the graphic below might explain this angular difference between Magnetic North and True North.

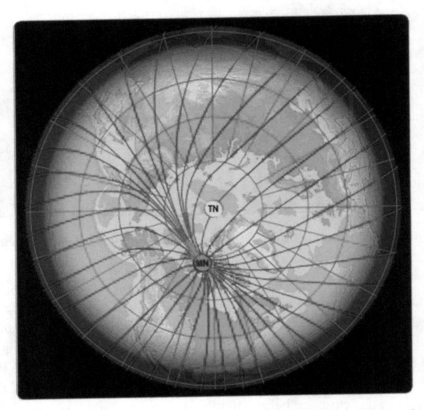

Figure 16-8. *Magnetic meridians are in red while the lines of longitude and latitude are in blue. From these lines of variation (magnetic meridians), one can determine the effect of local magnetic variations on a magnetic compass.*

I have a cute way to remember Magnetic Variation (I use this for school kids, but it works for adults too): Santa Claus lives at the "True North" Pole, but "Magnetic North" is hundreds of miles away from "True North." The difference between the two is what makes up the (magenta) lines of variation. Hope you can use it.

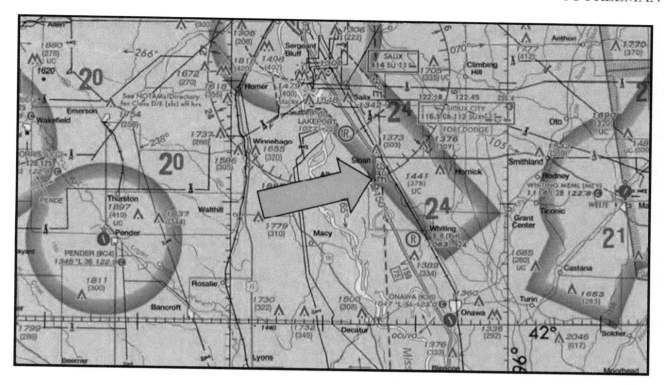

Figure 26, Magnetic Line of Variation (Source: FAA). Not to scale.

The straight magenta dashed line, Figure 26, Orange Arrow, shows the Magnetic Line of Variation. Bonus Knowledge: 3°E stands for a conversion factor that is used to calculate Magnetic Course. Now on to modern technology, GPS.

Global Positioning System (GPS)

Global Positioning System (GPS) is a satellite based navigation system that relies on speed of light calculation and triangulation between multiple satellites to accomplish very precise position finding. Here are some highlighted facts about GPS:

*24 Satellites (2 dozen) make up the system. I understand there a few spares but for FAA purposes 24 is the magic number.

*5 Satellites (just shy of a half dozen) are required to have signal to a GPS unit for full capability.

GPS technology is an awesome tool, but don't forget about how to find your way using lat/longs on a hard copy aeronautical chart.

Keep Studying Airspace

We suggest you take this airspace course from the FAA Safety Team (FAAST) website at: https://www.faasafety.gov/gslac/ALC/course_content.aspx?cID=42&sID=505&preview=true.

Practical Flight Planning Tip: You receive an aerial photography job request from the real estate representative that wants video of a newly constructed sub-division. Many people will use online driving map programs to verify the location and basic hazard information of the job site. These same sites will give you the latitude and longitude of the proposed site, which can be input into an online aeronautical chart program like Sky Vector to evaluate the local airspace for issues. We don't receive any compensation from Sky Vector but think they offer a very nice service.

Chapter 4

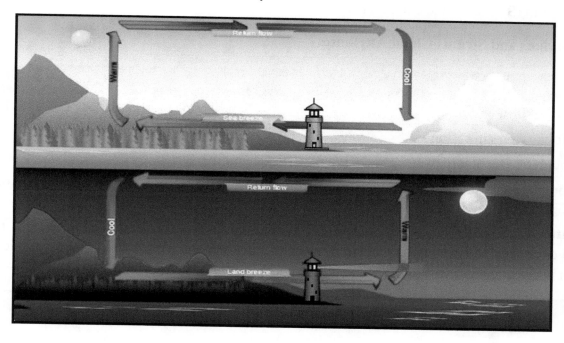

Aviation Weather: Basics, Resources and Effect on sUAS

If you were to ask me to define weather in ten words or less it would be, "the unequal heating of the Earth's surface." The Sun heats the various surfaces of the Earth differently. On a macro scale, the Ocean is heated differently than major land masses. On the micro scale, a parking lot is going to be heated differently than a lake. The byproducts of this unequal heating are; lifting action (instability) and humidity (water), the recipe for weather. The amount of unstable lifting action and humidity (water) available determines the vertical development, or height of the clouds. Here are some of the cloud types:

Cloud Types

Cumulous clouds are the puffy white clouds you see during clear days. However, with the right amount of lifting action, instability and humidity they can become powerful and dangerous storms. Inside of cumulous clouds are up and downdrafts as they build strength and cycling, water droplets until they reach the mature phase (more on this later) and let the water out as rain showers.

Stratus clouds have little vertical development (height), think of a calm winter day with an overcast layer of thin gray clouds. Stratus clouds are known for continuous precipitation (Memory aid: because of their coverage and consistency, they produce a uniform amount of rain). Since these clouds have less vertical development they are considered a stable air mass with is great turbulence-wise but not so good for visibility. See the attributes of each cloud type, referred to as Cumuliform (Cumulous) and Stratiform (Stratus) in the diagram below.

Cumulous and Stratus Cloud Attributes

Unstable Air	Stable Air
Cumuliform clouds	Stratiform clouds and fog
Showery precipitation	Continuous precipitation
Rough air (turbulence)	Smooth air
Good visibility (except in blowing obstructions)	Fair to poor visibility in haze and smoke

Lenticular, Rotor Clouds and Mountain Waves are known for dangerous turbulence (Mountain Waves) and clouds that produce strong winds being pushed over a mountain top (orographic lifting) as seen in the graphic below (a mountain wave condition minus rotor or lenticular clouds).

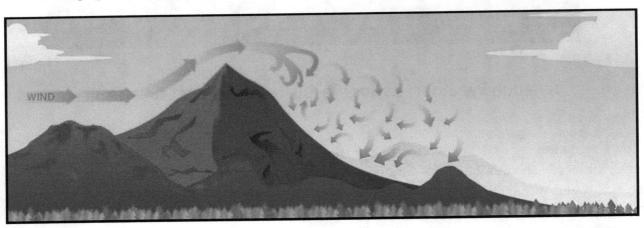

Mountain Wave Wind Turbulence (source: FAA)

The wind starts, on the left hand side of the graphic above, is called the Windward side of the mountain. As it crosses the mountain peak it is referred to as the Leeward side of the mountain. General orographic lifting clouds can form on the windward side of the mountain. As you might guess the leeward side of the mountain contains the dangerous turbulence with possible lenticular and rotor clouds (see below).

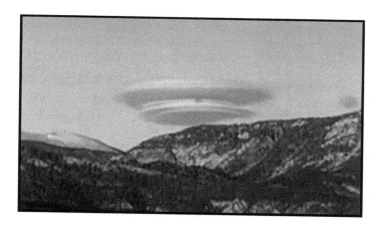

Lenticular Cloud (Source: Wikipedia, Marti8888, CC 4.0)

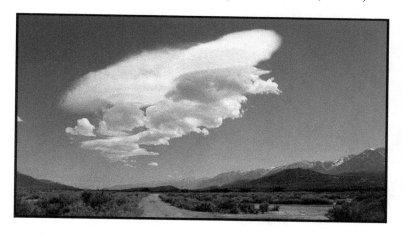

Rotor Clouds (Source: Wikipedia, CC 2.0, Malosse Flickrbot)

*There are a lot more cloud types out there you can research, but we are covering the main ones.

Thunderstorms

Thunderstorms are actually fully developed Cumulus clouds that deserve their own section. All Cumulus clouds start as the nice, fluffy, white and puffy clouds that you see on nice clear days (did I just say fluffy and puffy in the same sentence?). With the right amount of instability, lifting action and humidity, they can become Cumulonimbus (thunderstorm) clouds, the most feared clouds by pilots due to their destructive winds, dangerous turbulence and encompassing clouds that can cause the birth of neighboring thunderstorms called embedded thunderstorms. The nimbus in cumulonimbus means middle altitude clouds that range from 6,500' AGL to 40,000' AGL (and sometimes a little above). Thunderstorms will always have lightening (and thunder).

3 Stages of a Thunderstorm

There are **3 Stages of a Thunderstorm; Cumulus, Mature and Dissipating**.

***Cumulus** – The Cumulus stage is the white, puffy cloud stage, the cloud is building with up and down drafts.

***Mature** – At this stage the Cumulus cloud is building to its greatest size and strength, rain begins to fall this means the cloud cannot continue to hold the water droplets circulating in the up and down drafts within the storm cloud.

***Dissipating** – The Dissipating stage is the final stage in the thunderstorm process in which the downdrafts bring down the water contained in the clouds down to Earth in great volumes. After dissipating the water, the cloud dissipates as well. This process can be seen in graphic form below.

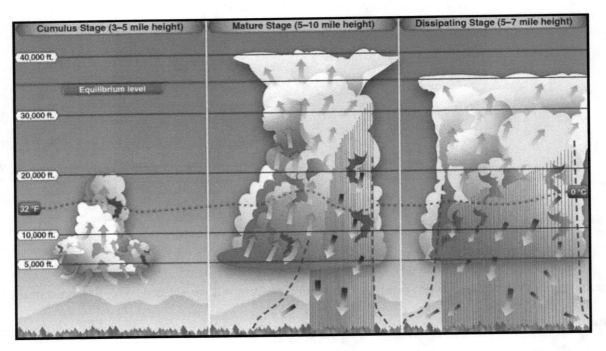

Life cycle of a Thunderstorm (source: FAA)

Temperature/Dew Point: A Quick Word

As we mentioned previously with regard to cloud development, it takes water (humidity) and instability (lifting action) to create clouds. It also takes the same for the creation of fog, which is essentially are clouds that rest on the surface of the ground (technically from the ground to 50' AGL). A gauge of the ideal conditions of visible moisture, be it clouds, fog, rain, thunderstorms, is the temperature/dew point. You have probably have heard this on the television evening news, weather segment, "temperature at this time is 70 degrees with a dew point of 60 degrees." Meteorologists have calculated that the ideal temperature based on the humidity and stability of the atmosphere that the dew point is 60° F. When the temperature gets within approximately 2 degrees of the dew point (62° or less), then some kind of visible moisture will be present. **The definition of temperature-dew point is the maximum amount**

of water vapor that air can hold and when water vapor condenses, (when it cannot hold it anymore); **it dissipates into visible moisture.** Excuse me; I need to run to the restroom.

Where does the atmosphere get its water to create visible moisture? It comes from evaporation and sublimation. Most people are familiar with evaporation, for example, lake water evaporating into the atmosphere. The creation of snow and ice are an example of sublimation, where water vapor converts directly to a solid bypassing the solid (water) phase.

All this talk about temperature and dew point is a good set-up for a conversation about fog, coming up after these messages (cut to commercial).

Fog

As we said before the fog is clouds that are at the surface (50' AGL or less above the ground). There are different types of fog based on either the temperature of the air or ground, bringing the temperature and dew point together. The warm ground can heat the air above or air flowing over cool water can cool the air. A practical application of using temperature and dew point is when you are on a mission site and there is fog. If you know the current temperature and dew point it is possible to roughly calculate when the mission can start based on the trend of rising temperature. The forecast with show this as well, but more on this later. Here are some of the types of fog.

*Steam Fog forms when cold air moves over warmer water, which can cause in the formation of ice (if below freezing) and low-level turbulence.

*Advection (movement of air) **Fog** happens when moist air moves over a cool surface. This happens in the winter when moist air blows in from the sea to the coast over cool ground.

*Radiation Fog occurs when the warmer water (or land) heats the cold air above its saturation (due) point producing fog. This situation is the most conducive to radiation fog is when cold air blows in over warm land.

*Upslope Fog is the movement of air upwards or upslope (orographic lifting). As temperature decreases to the dew point fog is created. There is Advection Upslope fog as well, translated, means, movement of air upslope.

Check out the picture below that nicely describes fog types:

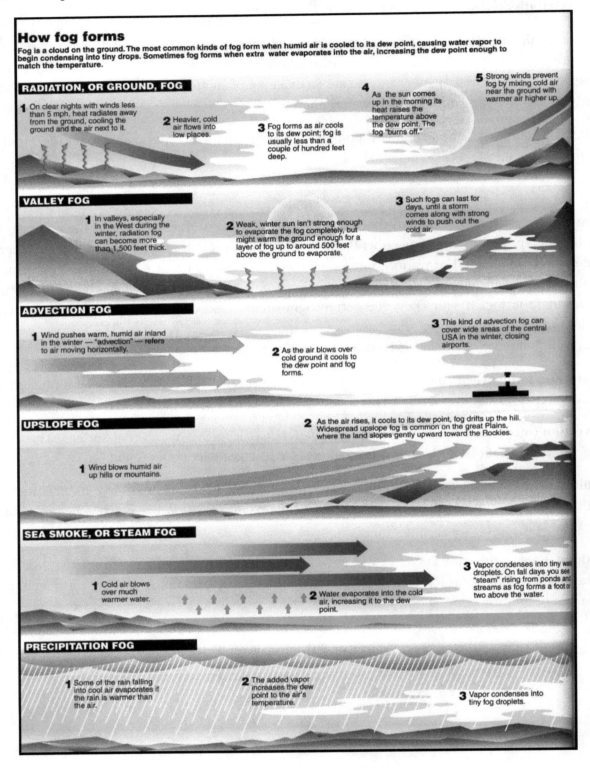

Types of Fog (Source: atomos.pccu.edu.tw)

Sea Breezes

Sea Breezes use a term called convective circulation to describe how they work. This is a very accurate description because the air traveling over cool water (cooling the air) moves inland and is then heated by the land travels up and back out to sea to start the process again in a circular pattern.

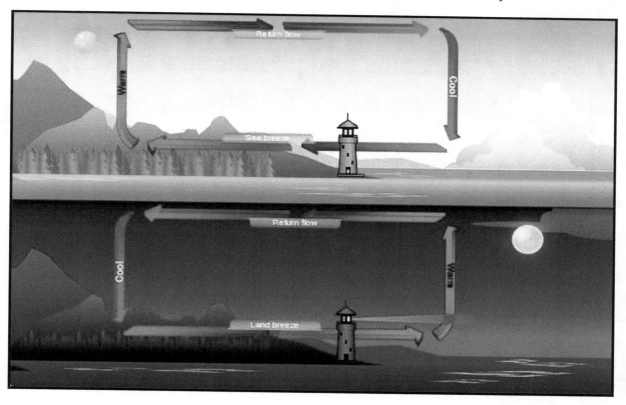

Land Breezes

This process happens at night when the landside cool air travels over the coast, is warmed, rises and returns back to the land in a circular cycle.

Micro Weather – Flying Over Various Surfaces

This chapter covers a lot of the high level weather information for larger areas, in other words, a macro scale but weather happens on a micro scale too. Flying over parking lots and plowed fields can create updrafts, while flying over water and low lying areas can cause downdrafts as shown in the graphic below.

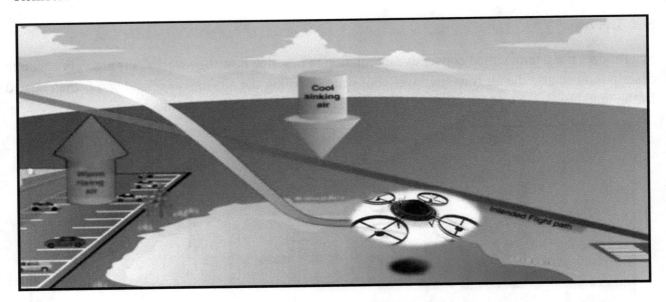

Turbulence from Buildings, Structures and Natural Obstructions

There have been many accidents with UA flying around structures, specifically large downtown skyscrapers (legality aside for the moment). The building themselves can cause turbulence as the wind works its way around obstacles or wind can channel in-between buildings (Bernoulli's Principal at work) increase speed. As a UA works between building the pilot isn't prepared for the associated wind and turbulence as illustrated in this manned aircraft example.

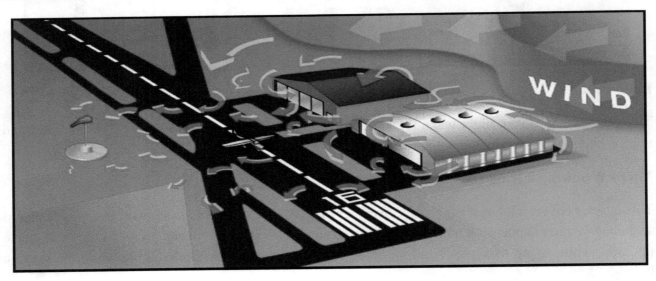

Inversions are temperature inversions, which can happen high level or low level. Air temperature tends to go down with increases in altitude (think of the temperature drop when driving up a mountain). As air pressure decreases with altitude, so does temperature. However, there are times when the opposite happens, air temperature increases with altitude. At low altitude, an inversion can act as a lid just over the surface, trapping pollution in a layer just above the ground.

Cloud Ceilings

As we work our way to taking our weather knowledge and applying it to reading weather reports we need to cover the concept of cloud ceilings. Folks not in aviation will look up in the sky, notice some clouds and exclaim, "cloudy day," without thinking about how high the clouds are above the ground. Weather reports contain information on cloud heights, it helps pilots know if they can safely make it to a destination. Clouds conditions are defined by sky coverage amounts and can be stated as; Clear (CLR), Sky Clear (SKC), Few (FEW), Scattered (SCT), Broken (BKN) or Overcast (OVC). These descriptions include a percentage of clouds covering the sky (increasing in coverage respectively, see graphic below) but only a BKN or OVC are considered official (cloud) ceilings by the FAA. This means that a pilot wanting to fly under VFR (relatively clear conditions) need only be concerned with BKN or OVC ceiling in-order to fly legally under VFR.

Cloud Sky Coverage

Sky Cover	Contraction
Less than ⅛ (Clear)	SKC, CLR, FEW
⅛–²⁄₈ (Few)	FEW
⅜–⁴⁄₈ (Scattered)	SCT
⅝–⅞ (Broken)	BKN
⁸⁄₈ or (Overcast)	OVC

Cloud Sky Coverage Chart (source: FAA)

Fronts

Fronts are familiar with most people that watch television weather. The meteorologist stands in front of the "green screen" showing how the various fronts are affecting local weather, not a bad place to start for pilots. Fronts describe areas of similar barometric and temperatures.

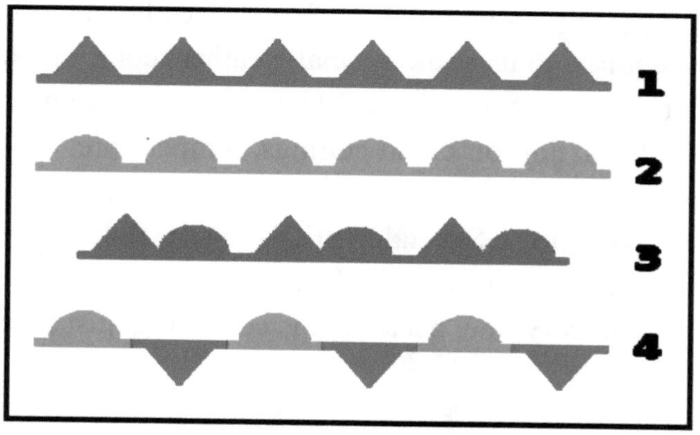

Types of Fronts (source: Wikipedia, -xfi-)

Fronts come in different varieties such as:

1) Cold Front – As the name suggests it is a colder air mass that is taking over a warmer air mass.

2) Warm Front – An air mass of warm air coming into contact with an area of colder air. Since the warm air is less dense, it tends to rise over the colder air.

3) Stationary Front – This type of front, is a kind of "stand-off," where neither air mass is taking over the other.

4) Occluded Front – Happens when a cold front overtakes a warm front, but is a conglomeration of three different air masses. There is also such a thing as a warm front occlusion but a cold front occlusion is more common.

Tip: The space between thunderstorms is called a front. Think of the frontal system (thunderstorm) activity.

This information will help you to start reading official government approved text weather briefings via the Automated Flight Service Station (www.1800WXBRIEF.com), so let us jump right in.

Reading Government Issued Weather (WX) Reports

In the modern age of computers, you would think that we would not still be producing weather reports the same way we did decades ago. This is when teletype machines only allowed a certain amount of character data to be transmitted at one time, enters weather shorthand. I guess you might say it is *Twitter* for the modern day with only 180 characters in which to broadcast weather information to the world. Decoding aviation weather can be frustrating at first, but keep practicing and it gets easier. First up is the backbone of meteorological reports METAR and TAFs.

METARs

At the core of reading aviation weather are Aviation Routine Weather Reports (METARs), I know the acronym doesn't make any sense, but is straight from the horse's mouth (http://www.faa.gov/jobs/abbreviations/). A METAR is a weather report regarding the immediate airport area. The airport is required to have the special weather reporting equipment to be included as official weather information. Reports are uploaded to the internet as an important part of a weather briefing. The report gives information on cloud ceilings, temperature, dew point, winds, barometric pressure and more. METAR reports are published every hour on the hour unless there is a significant change to the weather; then it is called a Special (SPECI) report. Before we look at one, check out the chart below and become familiar with the weather descriptors and phenomena codes below. Don't pay too much attention to the top category line, because it is confusing, just focus on the abbreviations.

Cool Tip: Listen to a phone recording of an automated weather broadcast (METAR) at the home of the Wright Brothers First Flight Airport (KFFA) by calling, (252) 449-0698. You can also use this website to find all of the phone numbers in the US: https://www.faa.gov/air_traffic/weather/asos/?state=FL.

Qualifier		Weather Phenomena		
Intensity or Proximity 1	Descriptor 2	Precipitation 3	Obscuration 4	Other 5
− Light	MI Shallow	DZ Drizzle	BR Mist	PO Dust/sand whirls
Moderate (no qualifier)	BC Patches	RA Rain	FG Fog	SQ Squalls
+ Heavy	DR Low drifting	SN Snow	FU Smoke	FC Funnel cloud
VC in the vicinity	BL Blowing	SG Snow grains	DU Dust	+FC Tornado or waterspout
	SH Showers	IC Ice crystals (diamond dust)	SA Sand	SS Sandstorm
	TS Thunderstorms	PL Ice pellets	HZ Haze	DS Dust storm
	FZ Freezing	GR Hail	PY Spray	
	PR Partial	GS Small hail or snow pellets	VA Volcanic ash	
		UP *Unknown precipitation		

The weather groups are constructed by considering columns 1–5 in this table in sequence: intensity, followed by descriptor, followed by weather phenomena (e.g., heavy rain showers(s) is coded as +SHRA).
* Automated stations only

Weather Descriptors (source: FAA)

There are many descriptors to learn and some are rarely used, but it might be on the test. In practical application, I would have this sheet printed for reference. Now for a METAR:

METAR KINK 121945Z 11012G19KT 15SM SKC 25/17 A3000

METAR KBOI 121854Z 13004KT 30SM SCT150 17/6 A3015

METAR KLAX 121852Z 25004KT 6SM BR SCT007 SCT250 16/15 A2991

SPECI KMDW 121856Z 32005KT 1 1/2SM RA OVC007 17/16 A2980 RMK RAB35

SPECI KJFK 121853Z 18004KT 1/2SM FG R04/2200 OVC005 20/18 A3006

Figure 12.—Aviation Routine Weather Reports (METAR).

Look at the first line, METAR for the KINK Airport, trust me, you will become familiar with the format by practicing. 121945Z means, the report was created on the 12th of the month at 1945 Zulu (time). 11012G19KT: Winds are from 110° (from the East-Southeast) at 12 Knots Gusting to 19 Knots. 15SM: Visibility 15 Statute Miles (SM). SKC: Sky Clear. 25/17: Temperature 25° C, Dew Point 17°. Barometric Pressure 30.00 Millibars.

Practice a few here, but you will find these in the Q & A section of this book.

TAFs

This acronym makes a lot more sense, TAF, Terminal Areas Forecast. TAFs are forecast covering a five-mile radius around an airport. The reports cover a 24 or 30-hour time period and are updated four times a day at 0000Z, 0600Z, 1200Z, and 1800Z. TAFs look a bit more daunting because of the massive amount of text, but the format is very familiar thanks to the METAR. Without further a-do…TAFs.

```
TAF

KMEM 121720Z 1218/1324 20012KT 5SM HZ BKN030 PROB40 2022 1SM TSRA OVC008CB
     FM2200 33015G20KT P6SM BKN015 OVC025 PROB40 2202 3SM SHRA
     FM0200 35012KT OVC008 PROB40 0205 2SM-RASN BECMG 0608 02008KT BKN012
     BECMG 1310/1312 00000KT 3SM BR SKC TEMPO 1212/1214 1/2SM FG
     FM131600 VRB06KT P6SM SKC=

KOKC 051130Z 0512/0618 14008KT 5SM BR BKN030 TEMPO 0513/0516 1 1/2SM BR
     FM051600 18010KT P6SM SKC BECMG 0522/0524 20013G20KT 4SM SHRA OVC020
     PROB40 0600/0606 2SM TSRA OVC008CB BECMG 0606/0608 21015KT P6SM SCT040=
```

FIGURE 15.—Terminal Aerodrome Forecasts (TAF).

Check out the KOKC TAF: The TAF was made on the 5th of the month at 1130 Z, it covers the dates/times from 12 Z on the 5th of the month to 18 Z on the 6th of the month. Winds are from 140° (from the Southeast) at 8 knots, 5 Statute Miles (SM) visibility in Mist (BR), clouds are, a Broken (BKN) layer at 3,000' AGL. Temporarily (TEMPO) between 13 Z and 16 Z on the 5th of the month visibility is expected to be 1½ SM in Mist (BR). From (FM) 16 Z on the 5th of the month winds are forecast to be from 180° (South) at 10 knots (KT), visibility greater (P) and 6 SM with Clear Skies (SKC) Becoming (BECMG) from 22 Z to 24 Z on the 5th of the month, winds 200° at 13 knots Gusting (G) to 20 knots, 4 SM visibility with Rain Showers (SHRA), Overcast (OVC) cloud ceiling. Probability (PROB) at 40% (40) that from 00 Z to 06 Z on the 6th of the month, visibility 2 SM, Thunderstorms and Rain Showers (TSRA), Overcast (OVC) cloud layer at 800' Cumulus (008CB) Becoming (BECMG) between 06 Z and 08 Z on the 6th of the month (0606/0608), winds from 210° at 15 knots (21015KT), visibility greater than 6 SM, Scattered Clouds (SCT) at 4,000', end of the report (=).

I know it can be overwhelming, but dive into the Q & A section to learn more. Here is a practical application tool I really dig.

Helicopter Emergency Medical Services (HEMS) Tool

One of my favorite tools for checking airport weather is the Helicopter Emergency Medical Services (HEMS) Tool, https://new.aviationweather.gov/hems. The tool was designed specifically for the emergency response community where low level flying is the norm. Weather is shown in an easy map format with information right at the pilot's fingertips from ground level to 5,000 feet. The convenience of this format is a time saver.

AIRMETs and SIGMETs are used to define exceptional (not a good exceptional) weather events. AIRMETs cover less severe forms of weather while SIGMETs cover SIGnificant weather issues.

AIRMETs cover less severe weather like; moderate turbulence and icing, sustained surface winds of 30 knots or more, or widespread restricted visibility.

SIGMETs give pilots information about **severe** turbulence and icing in addition to weather phenomenon such as; sandstorms, dust-storms and volcanic ash episodes that are reported as limiting visibility to 3 SM or less.

Convective SIGMETS includes severe thunderstorms, embedded thunderstorms, have dense coverage within an area or along a line. Thunderstorms must be 60 miles long with thunderstorms consisting of 40% of the area. Thunderstorms affecting a 3,000-mile area with 40% of the area affected. Thunderstorms that are embedded or severe forecast for more than 30 minutes during the valid period, regardless of the size of the area.

NOTAMs are (generally) information that is not shown on current charts, instrument approach plates (for pilots flying instrument approaches to an airport), etc. D NOTAMs consist of information regarding closures/construction at the airport. For the UA pilot it also includes uncharted obstructions, service notifications (like tower closure). Additionally, event information such as airshows and fly-ins are considered D NOTAMs. The Flight Data Center (FDC) NOTAMs cover changes not reflected in charts with respect to the airspace, instrument approaches (changes to the approach itself, not just obstructions and closures like with D NOTAMs) and airways.

A good portion of NOTAMs cover items temporary in nature like; Temporary Flight Restrictions (TFRs), large scale events, charted towers and more. This document published every 28 days, in the Notice to Airmen Publication (NTAP). Check out NTAP at; http://www.faa.gov/air_traffic/publications/notices/.

A weather briefer many include NOTAMs with the briefing, but Flight Data Center (FDC) NOTAMs are not covered unless specifically requested. FDC NOTAMs cover changes not reflected in charts with respect to the airspace, instrument approaches (changes to the approach itself, not just obstructions and closures like with D NOTAMs) and airways. If not given by the briefer you should also ask for D NOTAMs.

Practical Application: We suggest asking the briefer, "are there any NOTAMs, FDC or D NOTAMs, that will affect the area of my mission?" Some suggest that using the online briefing as an easier way to scan the data for items that may affect a flight operation due to the large quantities of Notice to Airmen Publication (NTAP) data. Because this information can be a huge amount of data it might be easier to consider looking this information online (www.1800WXBRIEF.com) to scan more efficiently the information for NOTAMs that may pertain to your flight.

All is Not Lost!

Go to NOAA Aviation Weather website and practice reading METARs, TAFs and a lot more; http://aviationweather.gov/. Click on the METARs tab on the first page, scroll down to this part for page:

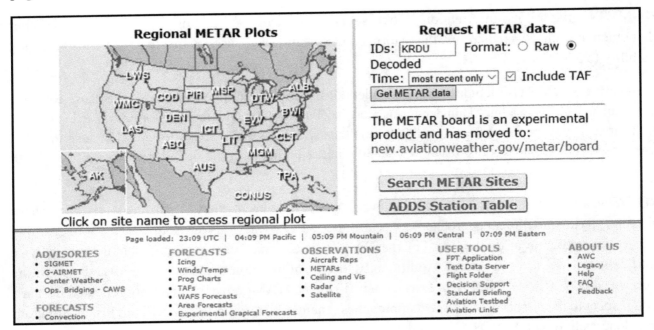

AviationWeather.gov website

Don't hate me for telling you this, but look closely at "format" section and you will see a choice between "raw" data and "decoded" (I know the layout is less than to be desired). Yes, it's true, you can decode all of those pesky weather reports. However, this won't keep you from needing learning the "raw" format of the test, but can be used as a tool for practicing.

Chapter 5

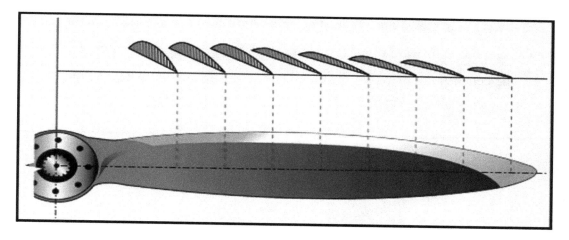

Flight Performance: planning, accessories and loads

Flight performance for manned aircraft includes; take-off and landing distance, climb performance, weight and balance (Center of Gravity calculations) and fuel burn. Depending on the manufacturer a sUAS pilot may or may not have this type of information available. If the information exists, the FAA requires you to use manufacturer provided data. In the absence of flight performance data, the Remote Pilot (RP) should search out calculations, others have made or use best practices of the industry.

Take-off and landing performance data can help the RP determine how much distance an aircraft will need to take-off considering the aircraft weight and temperature. A heavily loaded aircraft flown in high, humid temperature conditions will have greater take-off and landing distances in addition to poor climb performance.

Density Altitude is the effect of temperature and elevation on aircraft performance. An aircraft mission from a high mountaintop on a high, humid temperature day will perform more poorly and an aircraft flying in cold temperatures at a lower altitude on a low humidity day. Memory aid: High temperature, altitudes and humidity means air molecules are further apart, thus the propeller and wing has less "bite" on the air molecules. Combine the previously outlined conditions with a heavily loaded aircraft and performance is greatly degraded.

Center of Gravity (CG)

Center of Gravity (CG) is the balance point for the aircraft. Some sUAS manufactures have limits to how far the CG can move forward or aft (or from a center point with multi-rotor aircraft). The Center of Lift (CL) is the center point of where the culmination of lift from the wing originates. CL works in a symbiotic relationship with the CG.

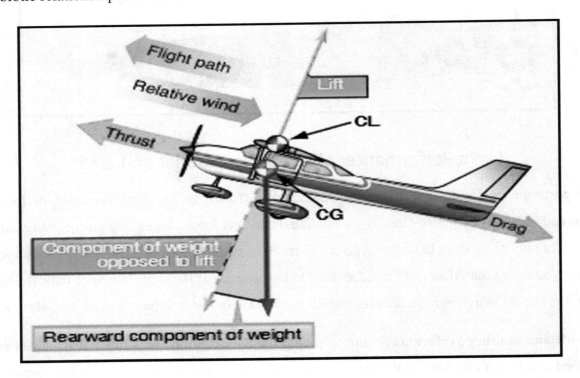

Center of Gravity (Source: FAA)

One of the major concerns with CG and sUAS is the effect of adding accessories (weight) to the aircraft. Put too much weight on the aircraft and flight performance can be greatly impacted. Alternatively, dropping an item from the aircraft (when permitted) will cause the aircraft to fly in a much lighter condition and may also radically alter the center of gravity, creating control issues. If the pilot isn't prepared for these type scenarios the aircraft can crash.

The addition of accessories, for example a cinematic quality camera lens and body, to the body of the aircraft will change not only the weight but the CG as well. Think of holding a 10 lb. weight away from your body, it seems heavier than holding close to your body. The weight effect of having a lever arm (of your arm) away from your body makes it seem heavier. In aircraft, there is only so much weight and distance from the normal CG you can place weight before having control issues with the aircraft.

sUAS owners must go by manufacturers guidelines, or industry best practices in the absence of manufacturer information, when calculating aircraft performance and weight and balance.

Chapter 6

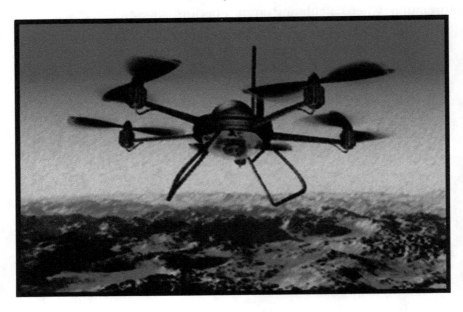

It Happens: Emergency Procedures

The regulations state that **a Remote Pilot (RP) can deviate from any regulation in the case of the emergency.** Regulations that are broken during the emergency should be just **to the point of being able to meet the need of the emergency** and nothing more. There's an old saying in the pilot community, "A pilot-in-command can break any rule in the event of an emergency, but there just might be a lot of paperwork to handle on the backside." In a nutshell, if you break a regulation, the FAA (Administrator) **may** ask you to submit a written report. There is a way to protect yourself from FAA enforcement, and you don't even need an attorney, by submitting a NASA Form very soon after the occurrence (more on this below).

When an Accident Must Be Reported According to the FAA (AC 107-2, 4.5)

The FAA classifies accidents that are required to be reported with standards based on injury and property damage (interestingly the property damage thresholds don't count your aircraft).

*Property Damage Thresholds

Property damage amounts to be classified as an accident is outlined in AC 107-2, 4.5:

Damage to any property, <u>other than the small UA,</u> if the cost is greater than $500 to repair or replace the property (whichever is lower).

Note: For example, a small UA damages a property whose fair market value is $200, and it would cost $600 to repair the damage. Because the fair market value is below $500, this accident is not required to be reported. Similarly, if the aircraft causes $200 worth of damage to property whose fair market value is $600, that accident is also not required to be reported because the repair cost is below $500.

*Injury Threshold

To be deemed an accident in the eyes of the FAA injury has to qualify as serious (Level 3 or greater) and per AC 107-2, 4.5:

*At least serious injury to any person or **any loss of consciousness**. A serious injury is an injury that **qualifies as Level 3 or higher** on the Abbreviated Injury Scale (AIS) of the Association for the Advancement of Automotive Medicine (AAAM). The AIS is an anatomical scoring system that provides a means of ranking the severity of an injury and is widely used by emergency medical personnel. Within the AIS system, injuries are ranked on a scale of 1 to 6, with Level 1 being a minor injury, Level 2 is moderate, Level 3 is serious, Level 4 is severe, Level 5 is critical, and Level 6 is a non-survivable injury. The FAA currently uses serious injury (AIS Level 3) as an injury threshold in other FAA regulations.*

Note: It would be considered a "serious injury" if a person requires hospitalization, but the injury is fully reversible (including, but not limited to, head trauma, broken bone(s), or laceration(s) to the skin that requires suturing).

Reporting to the FAA

If, according to the above information, the accident is classified as an accident, it is required to be reported to the appropriate **FAA Regional Operation Center within 10 Calendar Days**.

The report may (I don't think the FAA means "may" here. I would go with "must") *be submitted to the appropriate FAA Regional Operations Center (ROC) electronically or by telephone.* (If by phone I would insist on some proof, like an email, that the report has been received.)

Electronic reporting can be completed at www.faa.gov/uas/. To make a report by phone,

see Figure 4-1, FAA Regional Operations Centers Telephone List. Reports may also be

made to the nearest jurisdictional FSDO

(http://www.faa.gov/about/office_org/field_offices/fsdo/). The report should include the

following information:

1. sUAS remote PIC's name and contact information;

2. sUAS remote PIC's FAA airman certificate number;

3. sUAS registration number issued to the aircraft, if required (FAA registration

number);

4. Location of the accident;

5. Date of the accident;

6. Time of the accident;

7. Person(s) injured and extent of injury, if any or known;

8. Property damaged and extent of damage, if any or known; and

9. Description of what happened.

Reporting of some accidents to the NTSB AC 107-2, 4.5.2

Manned aircraft accidents are required to be reported to the NTSB. The FAA has agreed to handle the investigation of sUAS accidents (including initial reporting), however, this statement appears in AC 107-2:

4.5.2 www.ntsb.gov. National Transportation Safety Board (NTSB) Reporting. In addition to the report submitted to the ROC (FAA - Reginal Operations Center), and in accordance with the criteria established by the NTSB, certain sUAS accidents must also be reported to the NTSB. For more information, visit www.ntsb.gov.

Obviously this is a confusing procedure requiring the Remote Pilot (RP) to report accidents to the FAA and also the NTSB, which has their own criteria for accidents (primarily manned) covered in NTSB 830. My best hypothesis is that the NTSB might "pull rank" and decide to get involved if the accident if it is considered high profile. The only other scenario I can think of is if the accident involves a sUAS and a manned aircraft the NTSB would assume investigation responsibility of the entire investigation. My suggestion would be to report all accidents to both the FAA and NTSB, just in case. Don't assume that the FAA will automatically send the report to the NTSB for you.

Protect Your Remote Pilot Certificate Through the NASA Aviation Safety Reporting System (NASA ACRS)

Many years ago (1975) the FAA and NASA got together to figure out how to get more "in the field" information from pilots (including ATC, Aircraft Maintenance Technicians and others) about mistakes they made while flying. The problem is that pilots were reluctant to divulge this information fearing an FAA violation (called an Enforcement Action in Fed-speak). To solve this the FAA brought in NASA as an unbiased third party to handle reports from pilots. The next question was, "How to get pilots to use the system?" The answer was to offer a carrot of sorts by offering favorable treatment, if not entire dismissal, of any enforcement action the FAA may pursue in exchange for this information. Keep in mind, only a fraction of the pilot reported incidences are pursed by the FAA for enforcement action. Here are some highlights and tips for reporting:

* The violation must be reported immediately (technically you have up to 10 days) but you don't want to mess around with an enforcement action. Be sure to keep proof (either by electronic or hard copy, certified mail) that you sent the form.

* You can submit as many NASA Forms as you like, but can only use one every 7 years.

* Has to be non-criminal in nature. If it is, then you are on your own.

* No FAA enforcement action (violation) for a period of 5 years prior to the date of the occurrence.

Some pilots have affectionately called the NASA ASRS the "get out of jail free" card. While not to be taken literally, it can help produce a win-win situation in which pilots give information to the regulators, in a non-punitive fashion, that can ultimately enhance safety. Bookmark the NASA Form link on your computer and mobile device browser: http://asrs.arc.nasa.gov/report/electronic.html.

Chapter 7

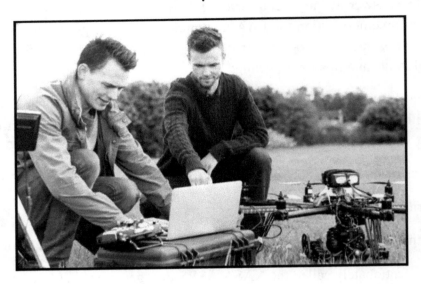

Aviation Communication for sUAS Flight Crews

Aviation communication is used to promote safety and efficiency while on a flight mission. Successful light crews can efficiently and effectively coordinate maneuvers and identify potential hazards. When operating near airports it is important that the UA crew listen to an aviation band radio to monitor area (aircraft) traffic and have an understanding of what they are doing. We recommend monitoring airports without an Air Traffic Control Tower (ATCT) to first to learn the basics. Since there isn't ATC at non-towered airports the job of listening and understanding is much easier. Non-towered airports use a specific, but open frequency of self-announce location (position) information. After you spend some time monitoring non-towered airports transition over to towered airports. You don't have to be in the immediate areas with an aviation band radio either, you can listen online at; http://www.liveatc.net/.

Airport Traffic Patterns

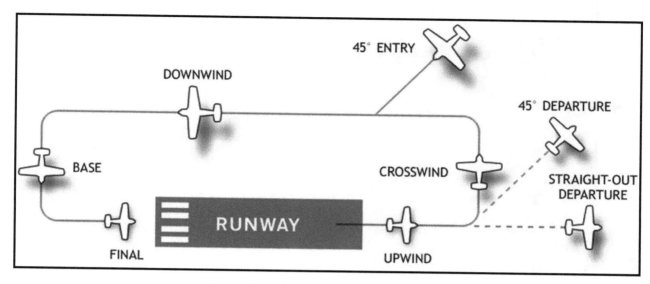

Airport Traffic Pattern (Wikipedia CC)

Traffic patterns are used to safely and efficiently manage air traffic into and out of airports, especially those without an ATCT. If a pilot knows that everyone is using the same system of arrival it makes non-towered airport operations safer. ATCT Airports use ATC to vector (direct) aircraft to and from the airport. Towered airports may use certain segments of the traffic pattern, more times than not, they will vector each aircraft to the "active" (or in-use) runway. Except for the 45° entry point all traffic pattern turns are made to the left, unless otherwise stated on the aeronautical chart or Airport Supplement. Departures are flown either straight out from the runway or via a left (45°) turn-out. Depending on the aircraft type, traffic patterns are flown by manned aircraft at 800' – 1,000' AGL. Corporate aircraft may fly a little higher at 1,500' AGL.

Scanning Technique

Scanning for hazards such as obstacles and hazards is an important duty. The technique you use needs to be efficient and effective. The technique is as follows: scan in 30° segments by looking far away from the mission sight (spot 1), while still having a clear vision, then proceed to the right (spot 2). After the distant areas are scanned, then focus closer (spot 3) followed by a scan to the left (spot 4).

The heart of the concept is that the human eye is limited seeing hazards through quick eye-movement. By focusing on small 30° segments the human eye performs will more easily see potential hazards.

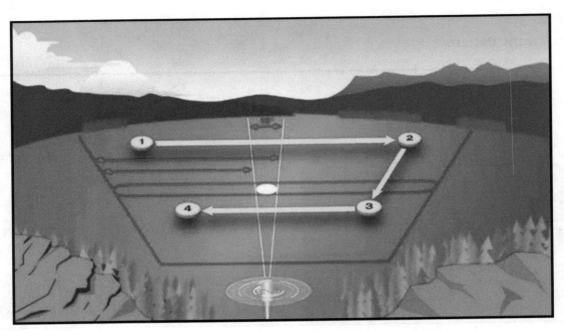

Recommended Scanning Technique (source: FAA)

The Remote Pilot (RP), Person Manipulating the Controls (PMC) and Visual Observer (VO) should develop a communication technique to point out hazards. We recommend using an adaptation of the clock reference technique manned aircraft pilots have used for decades. For the UA crew the PMC is the point of reference (like a manned aircraft pilot sitting in the aircraft) in which straight ahead is the 12 o'clock position. For instance, looking at the graphic above the PMC is located at the bottom of the V-shape with the UA (solid white circle) at the PMC's 12 o'clock position. 90° to the left is 9 o'clock and 90° to the left is 3 o'clock. An example of putting this to use would be when the VO notices a flock of large birds around Area 1 on the graphic and would state to the other crew members, "flock of birds, 10 to 11 o'clock position." The RP or PMC (depending on the prior agreed upon structure), "Roger, birds at 10 to 11 o'clock." Using this response ensures a message was communicated effectively (VO) and the message was received clearly and efficiently by the RP and PMC. Using this formal method might seem trivial to some, but if you become proficient at communicating with formal phraseology then the job of monitoring airport communication frequencies (Unicom at non-towered airports and Tower/Approach frequencies at ATCT airports) becomes easier.

Phonetic Alphabet

Character	Morse Code	Telephony	Phonic Pronunciation
A	• —	Alfa	(AL-FAH)
B	— • • •	Bravo	(BRAH-VOH)
C	— • — •	Charlie	(CHAR-LEE) or (SHAR-LEE)
D	— • •	Delta	(DELL-TAH)
E	•	Echo	(ECK-OH)
F	• • — •	Foxtrot	(FOKS-TROT)
G	— — •	Golf	(GOLF)
H	• • • •	Hotel	(HOH-TEL)
I	• •	India	(IN-DEE-AH)
J	• — — —	Juliett	(JEW-LEE-ETT)
K	— • —	Kilo	(KEY-LOH)
L	• — • •	Lima	(LEE-MAH)
M	— —	Mike	(MIKE)
N	— •	November	(NO-VEM-BER)
O	— — —	Oscar	(OSS-CAH)
P	• — — •	Papa	(PAH-PAH)
Q	— — • —	Quebec	(KEH-BECK)
R	• — •	Romeo	(ROW-ME-OH)
S	• • •	Sierra	(SEE-AIR-RAH)
T	—	Tango	(TANG-GO)
U	• • —	Uniform	(YOU-NEE-FORM) or (OO-NEE-FORM)
V	• • • —	Victor	(VIK-TAH)
W	• — —	Whiskey	(WISS-KEY)
X	— • • —	Xray	(ECKS-RAY)
Y	— • — —	Yankee	(YANG-KEY)
Z	— — • •	Zulu	(ZOO-LOO)
1	• — — — —	One	(WUN)
2	• • — — —	Two	(TOO)
3	• • • — —	Three	(TREE)
4	• • • • —	Four	(FOW-ER)
5	• • • • •	Five	(FIFE)
6	— • • • •	Six	(SIX)
7	— — • • •	Seven	(SEV-EN)
8	— — — • •	Eight	(AIT)
9	— — — — •	Nine	(NIN-ER)
0	— — — — —	Zero	(ZEE-RO)

Phonetic Alphabet

The phonetic alphabet is used to make communications clearer by assigning a word for each letter of the alphabet (see the figure above). Numbers are also handled in a way to prevent ambiguity, but are essentially the same with the exception of "tree for 3," "fife for 5"and "nine-er (nin-er) for 9." 3 and 5 are easily misunderstood, but 9 means "no" in German so they changed it to "nine-er." By the way, English is the International language of aviation. There are two great ways to learn the phonetic alphabet; one, listening to ATC and, two, practicing by picking out the letters of everyday items like street signs, signs and newspapers. For example, you see a street sign that states "Exit," and you would say, "Echo, X-Ray, India, Tango." If you practice out loud the folks around you might think you are mimicking a war movie actor calling out artillery locations. But seriously, the places you might use or hear the phonetic alphabet are with aircraft "tail numbers," for example a pilot may say, "November (November is a US aircraft), eight-one-seven-Foxtrot (N817F) is downwind Runway One Nine-er (Runway 19). Let us put this information to use by seeing some example of pilot communication at non-towered airports.

Non-Towered Airports (airport without an ATCT) Communications (Monitoring)

If you have a flight mission in the area of a non-towered airport, notify the airport about your flight mission. Once on the mission site monitor the airports frequency. This frequency, called a Unicom, and can be found on the aeronautical chart or in the Chart Supplement guide. Communication at non-towered airports are like an open CB Radio frequency (minus the extraneous conversation) in which pilots self-announce their position and intension over the communications (Unicom) frequency for which they are arriving or departing (or staying in the traffic pattern to practice takeoffs and landings). Here are some examples of non-towered airport communication:

"Town and Country traffic, Cessna 123 Bravo Foxtrot is 10 miles south inbound for landing, Town and Country traffic."

"Town and Country traffic, Cessna 123 Bravo Foxtrot, is entering the pattern, midfield left down-wind for runway 18, Town and Country traffic." *This means the pilot is entering the downwind leg of the traffic pattern using a 45° entry (see the airport traffic pattern graphic at the beginning of this chapter).

"Town and Country traffic, Cessna 123 Bravo Foxtrot, final, runway 18, Town and Country traffic." *The pilot is on the Final leg portion of the traffic pattern.

Air Traffic Control Tower (ATCT) Communications (Monitoring)

Flying in the ATCT airspace (Class B, C and D Airspace) requires prior permission from ATC. Once permission is received you should monitor the applicable ATC frequency to keep good Situational Awareness (SA) about the flight environment around you (which is the definition of SA). If you are within air traffic controlled airspace you should monitor the airports Tower frequency. This frequency can be found on the aeronautical chart and in the Airport Supplement guide. Outside of the immediate airport area use the airports Approach/Departure frequency listed in the Airport Supplement guide. Since aircraft are at higher altitudes while using Approach/Departure it won't necessarily be applicable to your UA flight operation. The best way to become familiar with ATC and pilot transmissions is to listen to ATCT frequencies, but we will start with the following examples.

ATC: *"November tree-fife-four-tango (N354T), cleared for takeoff runway two-tree (Runway 23), turn to a heading of one-eight-zero (180°)."* N345T is approved for takeoff on Runway 23 and is instructed to turn left to a heading of South. Here is the response from the pilot in N345T:

N345T: *"Cleared for takeoff Runway two-tree, turn to a heading of one-eight-zero."* This is a good example of the **Challenge and Response method of communication** in which ATC gives the Challenge (instructions) and the pilot Responds by repeating back the instructions. Bonus: Challenge and Response can be used as an effective tool in relationships too. And you thought this was just about aviation communication. Time to fix the sink, got to go.

ATC: *"November eight-one-victor, cleared for landing runway five right."* The pilot of N681V has permission to land on Runway 5 (stands for a 050 degrees) on the right runway, which means there is a left runway too (or at large airports a center runway followed by a left). Notice how the tail number was shorted by ATC to the last three digits. After full tail number, introductions are made by the pilot and ATC, ATC will shorten the tail number for efficiency sake. After ATC shortens the tail number (called an N number) then the pilot can do so. Since you are a guest at the air traffic controller airport, it's the proper thing to do.

N6681V: *"Eight-one-victor, cleared to land runway 5 right."*

Aviation communication has two components, listening, to the appropriate aviation frequency and to the crew members, then responding by repeating back the message received. Of course, responding is only for fellow crew members.

Pre-flight Briefing and Post Flight Debriefing

Speaking of aviation communication, don't forget the important steps of pre-briefing the flight mission with your crew. In this briefing the Remote Pilot (RP) can cover the mission and any issues that might arise plus what to do in the case of an emergency plus a Risk Factors evaluation (covered in the ADM Chapter). This should be led by the RP but also a time for the team to voice concerns and questions.

Emergency briefs can cover such things as helping with injuries, securing the area, and Emergency Response assistance. It's a good idea to have a fire extinguisher readily available and line tape for securing an area.

Debriefing is just as important as emergency and pre-briefing. After the mission is over the crew should discuss what went right and what went wrong, or could be improved upon, during the mission. One common failure of debriefings is that the findings are not acted upon or related to other flight crews as "lessons learned." Having this type of transparency will ultimately make the operation safer and more efficient.

Chapter 8

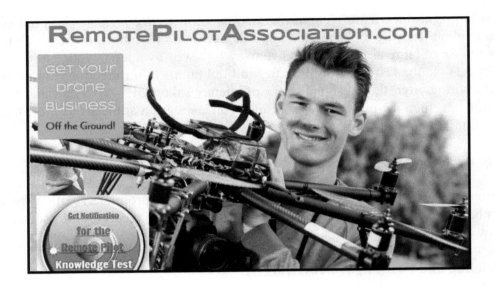

Fit to Fly

Every pilot has to be medically fit to fly in-order to legally operate a UA in the United States. Airline Transport Pilots working for an airline are under strict requirements, having to obtain a 1st Class Medical by visiting an Aero Medical Examiner (AMC), a physician with FAA authorization to administer the physical. On the other end of the spectrum glider and balloon pilots have to "self-certify" they are fit to fly. The tragic deaths of 16 people in the balloon accident in Kyle, Texas (2016) might bring the "self-certify" medical under scrutiny for commercial balloon operators and pilots. On another note, the Sport Pilot (license) certificate was created to encourage more people to learn how to fly small, simple aircraft, with the provision of being able to use their State's Driver's License for medical purposes. The saying is, "good enough to drive, good enough to fly." After years of data from the glider and Sport Pilot flying (with an ultra-low level accident rate due to medical issues, less than a fraction of 1%) the FAA, with a lot of encouragement from the pilot community, is going to extend the Driver's License Medical concept to aircraft weighing less than 6,000 lbs. (search 3rd Class medical reform for more information). The point of this history lesson is to point out that all pilots, regardless of the type of medical treatment they possess (or don't possess) are required to "self- certify" they are fit to fly before every flight. Remote Pilots should also take this philosophy back to their flight crews and impress upon them the responsibility of "self-certifying."

One of the often overlooked areas are, Over The Counter (OTC) drugs and prescription drugs, the most abused type of drugs in the Country. Most people think the words "under the influence" is referring to alcohol, but it also includes legal and illegal drugs. The difficult part of law enforcement is the testing for drugs, but that is a topic for another day. The best advice is to be aware of the side effects and warning associated with the OTC and prescription drugs you are taking. Antihistamines and decongestants are the two the main drugs that can cause issues with motor skills and cognitive processes (read the warning label).

Standards Around Alcohol

Alcohol is one of the substances that regulators have a lot of experience in dealing with regarding enforcement and have developed standards for operating motorized machinery. Aviation has the same rules for UA flight crews such as:

* Blood-alcohol concentration level is required to be less than 0.04.

* Alcohol can't be consumed within a window of 8 hours prior to flight. Memory aid: "8-hours bottle to throttle."

Note: You still can't be under the influence of alcohol, even after the 8-hour window has closed. I can specifically remember an incident in which an airline pilot was in court regarding an alcohol related incident. The pilot testified that he hadn't had any alcohol within 8 hours prior to the flight but had approximately a dozen mixed drinks just before the 8-hour cutoff. This is where the next rule caught him.

*Can't be under the influence of alcohol, drugs fall under this rule too.

Drugs

Not to belabor the point: As far as all drugs are concerned, I will quote the FAA, "…14 CFR part 107 and 14 CFR part 91, sections 91.17 and 91.19 prohibit the use of any drug that affects the person's faculties in any way contrary to safety. If you have a question to find out if a prescription drug is safe, ask an FAA AeroMedical Examiner (AME) or see FAA guidance to AMEs regarding common medications.

Remember to do research with herbal and dietary supplements, sports and energy boosters and products that contain the active components of some drugs.

I'M SAFE Checklist

```
┌─────────────────────────────────────────┐
│              I'M SAFE                     │
│              Checklist                    │
├───────────────────────────────────────────┤
│ Illness - Symptoms                        │
│ Medication - Prescription or OTC          │
│ Stress - Job, Financial, Health, Family   │
│ Alcohol - 8 Hrs? 24 Hrs?                  │
│ Fatigue - Adequately rested               │
│ Eating - Adequately Nourished             │
└───────────────────────────────────────────┘
```

The only issue I have with this particular I'M SAFE Checklist is that under "Alcohol" it states "24 Hrs." The rule is "8 hours bottle to throttle," so where does "24 Hrs." come into the picture? It might be a general guideline, for example, if you are going to a big party a day before the flight you should evaluate being able totally recover by the time of flight. Enough about this topic, let us move on to the excitable topic of hyperventilation.

Hyperventilation

Hyperventilation is caused by out of control breathing typically **brought on by a stressful situation** (or an emotionally charged situation). **Symptoms include; dizziness, lightheadedness, unconsciousness, visual impairment, hot and cold sensations, muscle spasms and tingling sensations.**

This fix to hyperventilation is **to bring Carbon Dioxide levels back to proper proportions** by **slowing the breathing rate**. Helping to get levels back into proportion can be aided by breathing in and out of a **paper bag** (I've used sick sacks because they are readily available in manned aircraft).

Stress

Stress comes in **2 types; Acute and Chronic. Acute Stress is the short term** stress most everyone encounters on a daily basis and **Chronic Stress is over a long period of time (the really bad type).**

Stress **releases Adrenaline** chemical hormones into the body. Stress as **negative effects; rise in blood pressure, blood sugar, heart rate, respiration and perspiration.** It can trigger a person's **"fight or flight"** mechanism in which we either fight our way out of a situation or extract ourselves away from it. Stress can be **classified as; physical stress (noise or vibration), physiological stress (fatigue) and psychological stress (difficult work or personal situations).**

Fatigue

Fatigue goes hand-in-hand with stress and are also categorized as **acute and chronic** that **affect motor skills and cognitive abilities.** There have been numerous studies proving that fatigue can have the same effect on a person as being impaired by alcohol. The worst type of fatigue result from; lack of sleep, over exercise and physical work.

TC's Two Cents: stress reactions have also made humanity survive for millions of years, it's the frequent and continuous type that is negative for the body and brain. If you find a way to eliminate all stress please let me know I have a cookie for you. The bottom line is that we should all be aware of our limits and actively seek out stress reduction techniques.

Dehydration

The elephant in the room with respect to health concerns with manned aircraft pilots is dehydration. The same can be said for **UA crews that work outside in a variety of environments.** In addition to having the proper tools and clothes for changing weather conditions, it is necessary to have plenty of water on-site for consumption. Notice I said water, **be careful of diuretic drinks; coffee, sodas, tea, etc.** It goes without saying, **dehydration can cause mental and physical impairment.** Drink at the first sign of thirst (if not before) and put plenty of water on your packing list.

Heat Exhaustion and Heat Stroke

Heat Exhaustion and Heat Stroke are heat stress conditions in **the body, maintaining a safe body temperature (heat exhaustion)** then progressing to the **body not being able to use the bodies**

internal temperature controls (heat stroke). Heat Exhaustion comes before Heat Stoke, both should be taken seriously, but Heat Stroke requires medical attention. When the body is over safe maximum temperature for too long it can damage internal organs.

Chapter 9

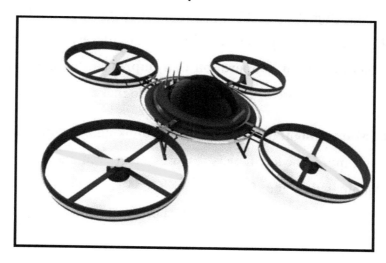

Guide to Better Decisions: Aeronautical Decision Making (ADM)

We have all watched television news report an aircraft accident when the reporter states the well-worn phrase, "the cause of the accident was believed to be pilot error." I remember Chelsey "Sully" Sullenberger saying in an interview one time something to the effect of, "if we don't look past the pilot in pilot error accidents we won't push the ball forward in safety." Unfortunately, 80% of accidents are attributed to pilot error (human factors), most being in the takeoff and landing portion (phase) of flight. **Aeronautical Decision Making (ADM) is defined as, "a systematic approach to risk assessment and stress management…(and) how personal attitudes affect decision-making (FAA)."** Over the past two-plus decades ADM has been positioned as the larger umbrella that many other risks and decision-making tools fall under. Some people will use Cockpit Resource Management and Human Factors inner-changeably but that's OK, just remember the FAA wants you to know ADM is where good decisions start.

Crew Resource Management (CRM)

Crew Resource Management (CRM), some say Cockpit Resource Management, got its start in the airlines in the 1960's. The airlines found that some of their pilots from the military (particularly solo flying military pilots) were having difficulty working as a crew after their military careers. Several airline accidents were attributed to a leadership style that was more hierarchical than team-like. Some Captains were known to be quite dogmatic when it came to their crew. **CRM is defined as a focus on the effective use of all available resources: human resources, hardware, and information supporting ADM to facilitate crew cooperation and improved decision-making.** The goal of all flight crews is good ADM and the use of CRM is one way to make good decisions.

Single-Pilot Resource Management (SRM) – If you are thinking, "I am just a guy with a quadcopter, what does CRM mean for me?" The key here is resources, and even the small time operator has them such as the; Pilot Manipulating the Controls (PMC), Visual Observer (VO), charts, Airport Supplements, Air Traffic Control, other area pilots, technology and more.

P.A.V.E.

PAVE is a tool that is used for risk management in flying by using an acronym that comprise all of the factors of risk for a flight in one place. I remembered this being kick-off fifteen years ago by John and Martha King, (King Schools) under contract with the FAA. Some human factors/CRM tools are less than be desired and, in my opinion, are heavy on the theory side of the equation and light on practical application. I have to say that PAVE has stood the test of time as a very useful tool I refer to frequently (and I am a hard sell). We will cover the PAVE acronym by describing each one:

*Pilot – Part of being a pilot is that we "self-certify" prior to each flight that we are mentally and physically fit to fly. Using the I'M SAFE Checklist can facilitate this process. Ask yourself, "Am I fit to fly?"

Example: The Pilot Manipulating the Controls (PMC) admitted on the flight mission site that is infant baby was up all night sick resulting in a severe lack of sleep. This can be a safety of flight issue.

*Aircraft – Your UA aircraft should be in a condition safety of flight called, airworthy. Flying with unresolved mechanical issues, and/or insidious quirks, raises your level of risk while on a flight mission.

Example: A job popped up last minute before mandatory software updates could be made to the Control Station (CS) computer. If you neglect to do the software update, this is a safety of flight issue.

*enVironment – All right, they used a little editorial creativity to make this acronym work but it's still useful. Evaluating your environment is being aware of the weather conditions, physical hazards and operational challenges that you face on a flight. This requires you to ask yourself, "is the level of risk is too high?"

Example: You notice on the mission site that there is a portion of the video shoot that will take you uncomfortably close to antenna guy wires. If the issue isn't discussed it could become a safety issue.

*External Pressures – I have to admit this is my personal favorite of the entire concept (forgive my "geek-out" statement). External pressures are the internal conversations to have with our self and with other people that negatively affect the safety of a flight. Examples of this would be time pressure to complete a job for a customer or something subtle like "cutting corners" due to company financial problems.

Example: A car lot owner demands that the video shoots not be cancelled again due to weather (because of a stationary front containing low visibility) because he has to get the footage to the local TV station for broadcast.

I could continue to "geek out" on this topic, but encourage you to think of some of your own examples from your business and personal life (I like those dual uses).

By keeping the PAVE tool close at hand before every flight will increase **Situational Awareness, the awareness and actions sensitive to all the elements of the PAVE principle.** When stress happens due to overload remember to stop, think, slow down and prioritize.

5 Hazardous Attitudes

The Five Hazardous Attitudes	Antidote
Anti-authority: "Don't tell me." This attitude is found in people who do not like anyone telling them what to do. In a sense, they are saying, "No one can tell me what to do." They may be resentful of having someone tell them what to do or may regard rules, regulations, and procedures as silly or unnecessary. However, it is always your prerogative to question authority if you feel it is in error.	Follow the rules. They are usually right.
Impulsivity: "Do it quickly." This is the attitude of people who frequently feel the need to do something, anything, immediately. They do not stop to think about what they are about to do, they do not select the best alternative, and they do the first thing that comes to mind.	Not so fast. Think first.
Invulnerability: "It won't happen to me." Many people falsely believe that accidents happen to others, but never to them. They know accidents can happen, and they know that anyone can be affected. However, they never really feel or believe that they will be personally involved. Pilots who think this way are more likely to take chances and increase risk.	It could happen to me.
Macho: "I can do it." Pilots who are always trying to prove that they are better than anyone else think, "I can do it—I'll show them." Pilots with this type of attitude will try to prove themselves by taking risks in order to impress others. While this pattern is thought to be a male characteristic, women are equally susceptible.	Taking chances is foolish.
Resignation: "What's the use?" Pilots who think, "What's the use?" do not see themselves as being able to make a great deal of difference in what happens to them. When things go well, the pilot is apt to think that it is good luck. When things go badly, the pilot may feel that someone is out to get them or attribute it to bad luck. The pilot will leave the action to others, for better or worse. Sometimes, such pilots will even go along with unreasonable requests just to be a "nice guy."	I'm not helpless. I can make a difference.

Read through the left hand side of the 5 Hazardous Attitudes chart below and try to think of an example in your own life, personal or professional, of people that have exhibited some (or all, hopefully not at one time) of these traits. You might see yourself in some of these examples. The 5 Hazardous Attitudes are; Anti-Authority, Impulsivity, Invulnerability, Macho, and Resignation.

After you have read through the left hand side of the graphic now transition to the right. DO IT NOW FOR CRYING OUT LOUD! Sorry, I'm being Impulsive. The right hand side is described as the Antidotes to the hazardous attitudes, which is the self-talk you might use for yourself to correct the situation. However, I like to use this as a tool while working with other people to evaluate safety being exhibited by individuals in a group. I have found that if a couple of people in an organization develop hazardous attitudes it can spread to the entire organization. But it is good to use as a self-evaluation tool as well.

Risk Management Decision-Making Process

Ugh, I have to admit this is my least favorite tools as a component of the ADM umbrella. A room full of PHD's probably worked overtime to come up with this gem. In a nutshell the process asks you to identify a problem (and its risk), make a decision, then evaluate that decision so the next time it happens you will be prepared how to handle it or have made changes to mitigate (or lessen the risk).

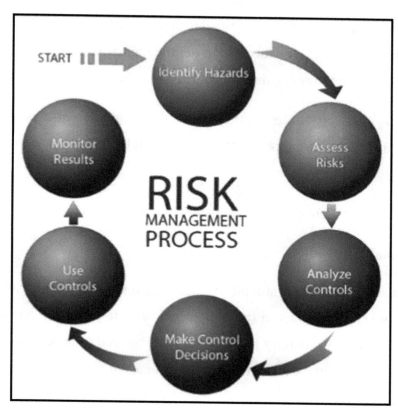

Risk Management Decision Making Process (source: FAA)

I guess we all will walk away with a fancy graphic.

3P Model of Decision Making – Yet another tool for better decision that uses the **3P Model; Perceive, Process and Perform**. Like the Risk Management Process, it consists of Perceiving, or identifying circumstances/issues related to flight, Process the information (evaluating risk) to come up with the solution and then Perform that solution (and evaluate).

CARE Decision Making Tool

Uses the acronym via a checklist format: **Consequences, Alternatives, Reality and External Factors.**

4 Ways of Handling Risk

Most people think of risk on one level by stating, "either it's risky or not." Risk isn't just risk alone, there are four actions you can take when developing an action surrounding risk, which is to: **Transfer, Eliminate, Accept, or Mitigate.**

***Transferring of Risk** is transferring it somewhere else, for example, a video shoot for an auto sales lot is in a confined shooting area with many obstacles. Instead of taking a high amount of risk with cast and crew members in the shot the crew decides to get the regular camera crew to shoot on top of the adjacent building for some of the shots, thus transferring the risk.

***Mitigating Risk** is to find ways to lessen the risk. If shooting video over people (waiver obtained) the flight crew finds a way to use the perimeter area of a flight path to avoid flying directly over the group.

***Accepting Risk** means the flight crew is aware and acknowledges the risk.

DECIDE Decision-making Model

The DECIDE Model is a recognized around the world as one of the standards of decision-making science. This acronym stands for: Detect (an issue), Estimate (if a decision has to be made), Choose (best decision for the outcome), Identify (what action are required to make the decision happen), Do (actions) and Evaluate (debrief the decision and make changes of next time).

The DECIDE model

1. **Detect.** The decision maker detects the fact that change has occurred.
2. **Estimate.** The decision maker estimates the need to counter or react to the change.
3. **Choose.** The decision maker chooses a desirable outcome (in terms of success) for the flight.
4. **Identify.** The decision maker identifies actions which could successfully control the change.
5. **Do.** The decision maker takes the necessary action.
6. **Evaluate.** The decision maker evaluates the effect(s) of his/her action countering the change.

DECIDE Decision-Making Tool (source: FAA)

One fact that has come to light is that the personalities and experience of the decision-maker come into play during decisions. Not all decisions are this well thought out, sent to committee (joke), in a process that cranks out the perfect decision. The DECIDE graphic above demonstrates two methods; **Analytical**, on the left, and **Automatic/Naturalistic**, on the right. There are many people (Ex. Miracle on the Hudson) that used the experience of the flight crew to make an immediate, and correct, decision (Automatic/Naturalistic) while others, like strategic planning, offers a whole host of other skill-sets (Analytical). We are not saying one is better than the other, just different.

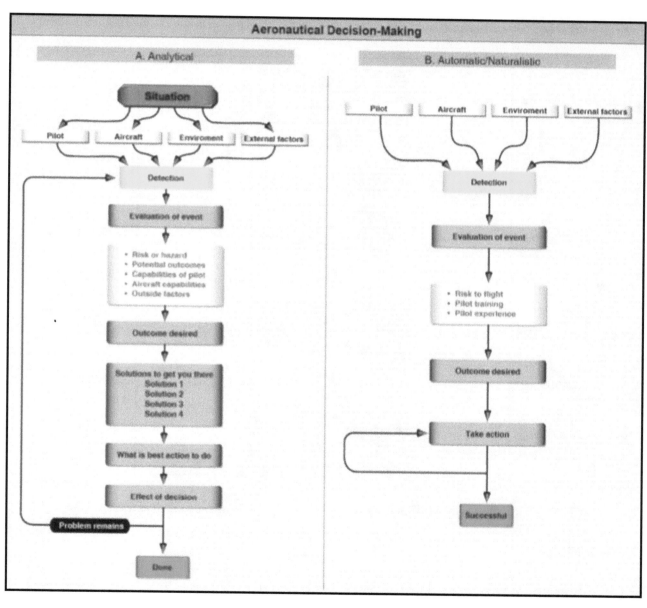

Decision-Making Analytical Verses Automatic/Naturalistic (source: FAA)

Don't be fooled, not making a decision is in-fact a decision. One thing I teach my flight students is that decisions have a time element associated with them. Some decisions require immediate action while others you have time in which to respond. The pilot's job is to triage which decisions have priority over others. In FAA-speak the process of **calculating priority along with a time element in decision making** (from immediate to long range planning) **process consist of; Strategic** (a lot of time to plan)**, Deliberate** (moderate time and planning) **and Time-Critical** (immediate based on the information at hand). The Miracle on the Hudson would be an example of a Time-Critical decision-making process.

Hazard and Risk Assessment Chart

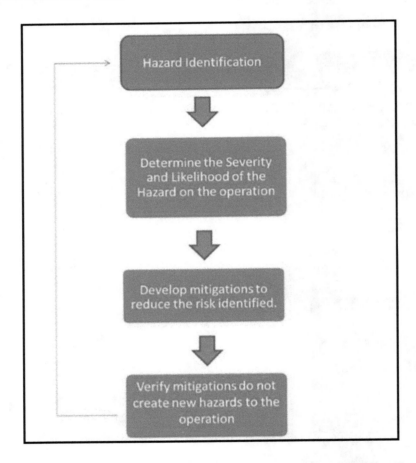

You might be thinking there are similarities with several of the graphics, and you would be right.

The Risk Matrix

Risk Analysis and Assessment. The risk assessment should use a conventional breakdown of risk by its **two components: likelihood of occurrence and severity**. Risk is a concept of the level of risk while also considering severity. The graphic below will assign a number level of severity and risk that will culminate to a total risk matrix.

Table A-1. Sample Severity and Likelihood Criteria

Severity of Consequences			Likelihood of Occurrence		
Severity Level	Definition	Value	Likelihood Level	Definition	Value
Catastrophic	Equipment destroyed, multiple deaths.	5	Frequent	Likely to occur many times	5
Hazardous	Large reduction in safety margins, physical distress, or a workload such that crewmembers cannot be relied upon to perform their tasks accurately or completely. Serious injury or death. Major equipment damage.	4	Occasional	Likely to occur sometimes	4
Major	Significant reduction in safety margins, reduction in the ability of crewmembers to cope with adverse operating conditions as a result of an increase in workload, or as result of conditions impairing their efficiency. Serious incident. Injury to persons.	3	Remote	Unlikely, but possible to occur	3
Minor	Nuisance. Operating limitations. Use of emergency procedures. Minor incident.	2	Improbable	Very unlikely to occur	2
Negligible	Little consequence.	1	Extremely Improbable	Almost inconceivable that the event will occur	1

The next step in the process is **consider the corresponding color code (see the graphic below); Green – Acceptable, Yellow – Acceptable with Mitigation and Red – Unacceptable.**

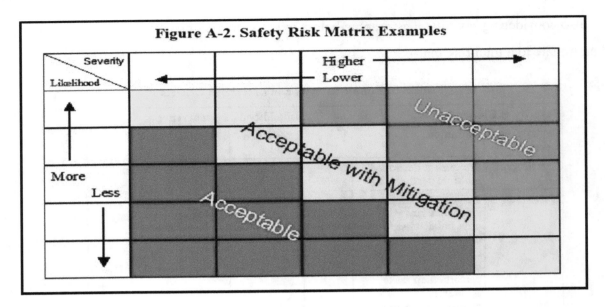

Safety Risk Matrix general concept.

Table A-2. Safety Risk Matrix—Example

Risk Likelihood		Risk Severity				
		Catastrophic A	Hazardous B	Major C	Minor D	Negligible E
Frequent	5	5A	5B	5C	5D	5E
Occasional	4	4A	4B	4C	4D	4E
Remote	3	3A	3B	3C	3D	3E
Improbable	2	2A	2B	2C	2D	2E
Extremely Improbable	1	1A	1B	1C	1D	1E

Note: The direction of higher/lower and more/less scales on a matrix is at the discretion of the remote PIC.

Safety Risk Matrix example

The Safety Risk Matrix graphic above is where we **put together the number value of risk with the color code.**

Color Yellow Risk Mitigation example: *An example of this situation would be an assessment of the impact of a sUAS operation near a school yard* and the flight team deems this risk a 4C using the matrix (an Occasional Risk rated 4 and Severity deemed Major C – 4C). *Scheduling the operation to take place when school is not in session could be one mitigation to prevent undue risk to the children that study and play there. Another mitigation could be restricting people from the area of operations by placing cones or security personnel to prevent unauthorized access during the sUAS flight operation.*

Where I can see the value in the Risk Matrix tool is being able to sit down outside of the flight environment and decide what type of risk is acceptable through scenarios, either through experience or through others.

The multiple tools provided in this chapter can you into a state of overwhelm. I think the key is to pick a couple of concepts you like and employ those techniques to your operation. One of the best ways to counteract making bad decisions (a.k.a. operational pitfalls) is to learn from others. Study of accident reports will allow you to see trends in human performance issues to see beyond pilot error and mitigate risk.

Chapter 10

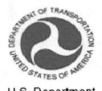

Code of Federal Regulations (CFRs)

U.S. Department
of Transportation

Federal Aviation
Administration

Brought to you by:
Federal Aviation Adminstration

Rules of the Air

Part of understanding the "rules of the air," or more correctly the Code of Federal Regulations (CFRs) is learning how to interpret the "legal-ese" fashion in which it is written. The FAA has gotten better about making regulations, or "regs" (slang for aviation CFRs) easier to read, starting with the Sport Pilot license (certificate) that came out several years ago. Tip: Read the entire regulation, including any cross referenced information. The entire interpretation of the regulation can change by excluding certain passages. Also pay attention to the words, "should, must, may and required," because these words will tip you off to the fact if the regulation is a "must" or "optional." Additionally, when the regulatory state, "or by the Administrator," don't worry, Michael Huerta, the FAA Administrator, is not going to show up at your house, it just means a representative from the FAA acting on behalf of the Administrator.

Advisory Circular (AC) 107-2

The new 14 CFR Part 107 is another step in the right direction. In this same theme the FAA has a document that goes into detail to explain frequently misunderstood CFR's called Advisory Circulars (ACs). The FAA's disclaimer is that ACs are not a legal interpretation, but rather an elaboration about CFRs already in existence. Read the document here: http://www.faa.gov/uas/media/AC_107-2_AFS-1_Signed.pdf. I know sometimes links are not taken seriously, but it's important to read this particular document in its entirety. However, we have included the actual regulation portion below:

14 CFR Part 107

PART 107–SMALL UNMANNED AIRCRAFT SYSTEMS

§ 107.71 Retesting after failure.
§ 107.73 Initial and recurrent knowledge test.
§ 107.74 Initial and recurrent training courses.
§ 107.77 Change of name or address.
§ 107.79 Voluntary surrender of certificate.
Subpart D—Waivers
§ 107.200 Waiver policy and requirements.
§ 107.205 List of regulations subject to waiver.

Authority: 49 U.S.C. 106 (f), 40101 note, 40103 (b), 44701 (a)(5); Sec. 333 of Pub. L. 11295

Subpart A—General§ 107.1 Applicability.

(a) Except as provided in paragraph (b) of this section, this part applies to the registration, airman certification, and operation of civil small unmanned aircraft systems within the United States.
(b) This part does not apply to the following:
(1) Air carrier operations;
(2) Any aircraft subject to the provisions of part 101 of this chapter; or
(3) Any operation that a remote pilot in command elects to conduct pursuant to an exemption issued under section 333 of Public Law 112-95, unless otherwise specified in the exemption.

§ 107.3 Definitions.

The following definitions apply to this part. If there is a conflict between the definitions of this part and definitions specified in § 1.1 of this chapter, the definitions in this part control for purposes of this part:

Control station means an interface used by the remote pilot to control the flight path of the small unmanned aircraft.

Corrective lenses means spectacles or contact lenses.

Small unmanned aircraft means an unmanned aircraft weighing less than 55 pounds at takeoff, including everything that is on board or otherwise attached to the aircraft.

Small unmanned aircraft system (small UAS) means a small unmanned aircraft and its associated elements (including communication links and the components that control the small unmanned aircraft) that are required for the safe and efficient operation of the small unmanned aircraft in the national airspace system.

<u>Unmanned aircraft</u> means an aircraft operated without the possibility of direct human intervention from within or on the aircraft.

<u>Visual observer</u> means a person who is designated by the remote pilot in command to assist the remote pilot in command and the person manipulating the flight controls of the small UAS to see and avoid other air traffic or objects aloft or on the ground.

§ 107.5 Falsification, reproduction, or alteration.

(a) No person may make or cause to be made

(1) Any fraudulent or intentionally false record or report that is required to be made, kept, or used to show compliance with any requirement under this part.

(2) Any reproduction or alteration, for fraudulent purpose, of any certificate, rating, authorization, record or report under this part.

(b) The commission by any person of an act prohibited under paragraph (a) of this section is a basis for any of the following:

(1) Denial of an application for a remote pilot certificate or a certificate of waiver,

(2) Suspension or revocation of any certificate or waiver issued by the Administrator under this part and held by that person; or

(3) A civil penalty.

§ 107.7 Inspection, testing, and demonstration of compliance.

(a) A remote pilot in command, owner, or person manipulating the flight controls of a small unmanned aircraft system must, upon request, make available to the Administrator:

(1) The remote pilot certificate with a small UAS rating; and

(2) Any other document, record, or report required to be kept under the regulations of this chapter.

(b) The remote pilot in command, visual observer, owner, operator, or person manipulating the flight controls of a small unmanned aircraft system must, upon request, allow the Administrator to make any test or inspection of the small unmanned aircraft system, the remote pilot in command, the person manipulating the flight controls of a small unmanned aircraft system, and, if applicable, the visual observer to determine compliance with this part.

§ 107.9 Accident reports.

No later than 10 calendar days after an operation that meets the criteria of either paragraph (a) or (b) of this section, a remote pilot in command must report to the FAA, in a manner acceptable to the Administrator, any operation of the small unmanned aircraft involving at least:

(a) Serious injury to any person or any loss of consciousness; or

(b) Damage to any property, other than the small unmanned aircraft, unless one of the following conditions is satisfied:

(1) The cost of repair (including materials and labor) does not exceed $500; or

(2) The fair market value of the property does not exceed $500 in the event of total loss.

Subpart B—Operating Rules
§ 107.11 Applicability.

This subpart applies to the operation of all civil small unmanned aircraft systems subject to this part.

§ 107.12 Requirement for a remote pilot certificate with a small UAS rating.

(a) Except as provided in paragraph (c) of this section, no person may manipulate the flight controls of a small unmanned aircraft system unless:

(1) That person has a remote pilot certificate with a small UAS rating issued pursuant to Subpart C of this part and satisfies the requirements of § 107.65; or

(2) That person is under the direct supervision of a remote pilot in command and the remote pilot in command has the ability to immediately take direct control of the flight of the small unmanned aircraft.

(b) Except as provided in paragraph (c) of this section, no person may act as a remote pilot in command unless that person has a remote pilot certificate with a small UAS rating issued pursuant to Subpart C of this part and satisfies the requirements of § 107.65.

(c) The Administrator may, consistent with international standards, authorize an airman to operate a civil foreign-registered small unmanned aircraft without an FAA-issued remote pilot certificate with a small UAS rating.

§ 107.13 Registration.

A person operating a civil small unmanned aircraft system for purposes of flight must comply with the provisions of § 91.203 (a) (2).

§ 107.15 Condition for safe operation.

(a) No person may operate a civil small unmanned aircraft system unless it is in a condition for safe operation. Prior to each flight, the remote pilot in command must check the small unmanned aircraft system to determine whether it is in a condition for safe operation.

(b) No person may continue flight of the small unmanned aircraft when he or she knows or has reason to know that the small unmanned aircraft system is no longer in a condition for safe operation.

§ 107.17 Medical condition.

No person may manipulate the flight controls of a small unmanned aircraft system or act as a remote pilot in command, visual observer, or direct participant in the operation of the small unmanned aircraft if he or she knows or has reason to know that he or she has a physical or mental condition that would interfere with the safe operation of the small unmanned aircraft system.

§ 107.19 Remote pilot in command.

(a) A remote pilot in command must be designated before or during the flight of the small unmanned aircraft.

(b) The remote pilot in command is directly responsible for and is the final authority as to the operation of the small unmanned aircraft system.

(c) The remote pilot in command must ensure that the small unmanned aircraft will pose no undue hazard to other people, other aircraft, or other property in the event of a loss of control of the aircraft for any reason.

(d) The remote pilot in command must ensure that the small UAS operation complies with all applicable regulations of this chapter.

(e) The remote pilot in command must have the ability to direct the small unmanned aircraft to ensure compliance with the applicable provisions of this chapter.

§ 107.21 In-flight emergency.

(a) In an in-flight emergency requiring immediate action, the remote pilot in command may deviate from any rule of this part to the extent necessary to meet that emergency.

(b) Each remote pilot in command who deviates from a rule under paragraph (a) of this section must, upon request of the Administrator, send a written report of that deviation to the Administrator.

§ 107.23 Hazardous operation.

No person may:

(a) Operate a small unmanned aircraft system in a careless or reckless manner so as to endanger the life or property of another; or

(b) Allow an object to be dropped from a small unmanned aircraft in a manner that creates an undue hazard to persons or property.

§ 107.25 Operation from a moving vehicle or aircraft.

No person may operate a small unmanned aircraft system -

(a) From a moving aircraft; or

(b) From a moving land or waterborne vehicle unless the small unmanned aircraft flies over a sparsely populated area and is not transporting another person's property for compensation or hire.

§ 107.27 Alcohol or drugs.

A person manipulating the flight controls of a small unmanned aircraft system or acting as a remote pilot in command or visual observer must comply with the provisions of §§ 91.17 and 91.19 of this chapter.

§ 107.29 Daylight operations.

(a) No person may operate a small unmanned aircraft system during the night.

(b) No person may operate a small unmanned aircraft system during periods of civil twilight unless the small unmanned aircraft has lighted the anti-collision lighting visible for at least 3 statute miles. The remote pilot in command may reduce the intensity of the anti-collision lighting if he or she determines that, because of operating conditions, it would be in the interest of safety to do so.

(c) For purposes of subsection (b) of this section, civil twilight refers to the following:

(1) Except for Alaska, a period of time that begins 30 minutes before official sunrise and ends at official sunrise;

(2) Except for Alaska, a period of time that begins at official sunset and ends 30 minutes after official sunset; and

(3) In Alaska, the period of civil twilight as defined in the Air Almanac.

This can be a tricky question with all of the "civil twilight" talk thrown in the mix. If you think about an average day, the Sun actually rises before official sunrise, this is civil twilight, and it starts: 30 minutes before official sunrise. As far as night is concerned notice how at official sunset there is still a good bit of light remaining, this time is civil twilight. You are approved to fly: 30 minutes after official sunset. This is, provided you have a lighting system that can be seen for 3 miles.

You might get a question such as; You arrive on the flight mission site: 40 minutes before sunrise, how long must you wait before you can begin flying (with an aircraft that has lighting that is visible for 3 miles)? Answer: 10 minutes, because you are: 10 minutes from the beginning of civil twilight. Be careful if you are asking the same question, but it doesn't say the aircraft has the approved lights.

§ 107.31 Visual line of sight aircraft operation.

(a) With vision that is unaided by any device other than corrective lenses, the remote pilot in command, the visual observer (if one is used), and the person manipulating the flight control of the small unmanned aircraft system must be able to see the unmanned aircraft throughout the entire flight in order to:

(1) Know the unmanned aircraft's location;

(2) Determine the unmanned aircraft's attitude, altitude, and direction of flight;

(3) Observe the airspace for other air traffic or hazards; and

(4) Determine that the unmanned aircraft does not endanger the life or property of another.

(b) Throughout the entire flight of the small unmanned aircraft, the ability described in subsection (a) of this section must be exercised by either:

(1) The remote pilot in command and the person manipulating the flight controls of the small unmanned aircraft system; or

(2) A visual observer.

§ 107.33 Visual observer.

If a visual observer is used during the aircraft operation, all of the following requirements must be met:

(a) The remote pilot in command, the person manipulating the flight controls of the small unmanned aircraft system, and the visual observer must maintain effective communication with each other at all times.

(b) The remote pilot in command must ensure that the visual observer is able to see the unmanned aircraft in the manner specified in § 107.31.

(c) The remote pilot in command, the person manipulating the flight controls of the small unmanned aircraft system, and the visual observer must coordinate to do the following:

(1) Scan the airspace where the small unmanned aircraft is operating, for any potential collision hazard; and

(2) Maintain awareness of the position of the small unmanned aircraft through direct visual observation.

§ 107.35 Operation of multiple small unmanned aircraft.

A person may not operate or act as a remote pilot in command or visual observer in the operation of more than one unmanned aircraft at the same time.

§ 107.36 Carriage of hazardous material.

A small unmanned aircraft may not carry hazardous material. For purposes of this section, the term hazardous material is defined in 49 CFR 171.8.

§ 107.37 Operation near aircraft; right-of-way rules.

(a) Each small unmanned aircraft must yield the right of way to all aircraft, airborne vehicles, and launch and reentry vehicles. Yielding the right of way means that the small unmanned aircraft must give way to the aircraft or vehicle and may not pass over, under, or ahead of it unless well clear.

(b) No person may operate a small unmanned aircraft so close to another aircraft as to create a collision hazard.

§ 107.39 Operation over human beings.

No person may operate a small unmanned aircraft over a human being unless that human being is:
    (a) Directly participating in the operation of the small unmanned aircraft; or
    (b) Located under a covered structure or inside a stationary vehicle that can provide reasonable protection from a falling small unmanned aircraft.

§ 107.41 Operation in certain airspace.

No person may operate a small unmanned aircraft in Class B, Class C, or Class D airspace or within the lateral boundaries of the surface area of Class E airspace designated for an airport unless that person has prior authorization from Air Traffic Control (ATC).

This is done through a request for a waiver on the main FAA website (and not with the individual tower) at: https://www.faa.gov/uas/request_waiver/.

§ 107.43 Operation in the vicinity of airports.

No person may operate a small unmanned aircraft in a manner that interferes with operations and traffic patterns at any airport, heliport, or seaplane base.

§ 107.45 Operation in prohibited or restricted areas.

No person may operate a small unmanned aircraft in prohibited or restricted areas unless that person has permission from the using or controlling agency, as appropriate.

The "controlling agency" can be found in the parameter of the chart, which also includes information such as times of serviced and altitudes covered.

§ 107.47 Flight restrictions in the proximity of certain areas designated by notice to airmen.

A person acting as a remote pilot in command must comply with the provisions of §§ 91.137 through 91.145 and 99.7 of this chapter.

§ 107.49 Preflight familiarization, inspection, and actions for aircraft operation.

Prior to flight, the remote pilot in command must:
    (a) Assess the operating environment, considering risks to persons and property in the immediate vicinity both on the surface and in the air. This assessment must include:
    (1) Local weather conditions;

(2) Local airspace and any flight restrictions;

(3) The location of persons and property on the surface; and

(4) Other ground hazards.

(b) Ensure that all persons directly participating in the small unmanned aircraft operation are informed about the operating conditions, emergency procedures, contingency procedures, roles and responsibilities, and potential hazards;

(c) Ensure that all control links between the ground control station and the small unmanned aircraft are working properly;

(d) If the small unmanned aircraft is powered, ensure that there is enough available power for the small unmanned aircraft system to operate for the intended operational time; and

(e) Ensure that any object attached or carried by the small unmanned aircraft is secure and does not adversely affect the flight characteristics or controllability of the aircraft.

§ 107.51 Operating limitations for small unmanned aircraft.

A remote pilot in command and the person manipulating the flight controls of the small unmanned aircraft system must comply with all of the following operating limitations when operating a small unmanned aircraft system:

(a) The ground speed of the small unmanned aircraft may not exceed 87 knots (100 miles per hour).

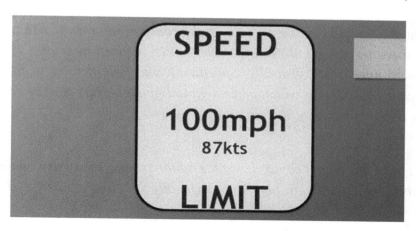

(b) The altitude of the small unmanned aircraft cannot be higher than 400 feet above ground level, unless the small unmanned aircraft:

(1) Is flown within a 400-foot radius of a structure; and

(2) Does not fly higher than 400 feet above the structure's immediate uppermost limit.

(c) The minimum flight visibility, as observed from the location of the control station must be no less than 3 statute miles. For purposes of this section, flight visibility means the average slant distance from the control station at which prominent unlighted objects may be seen and identified by day and prominent lighted objects may be seen and identified by night.

(d) The minimum distance of the small unmanned aircraft from clouds must be no less than:

(1) 500 feet below the cloud; and

(2) 2,000 feet horizontally from the cloud.

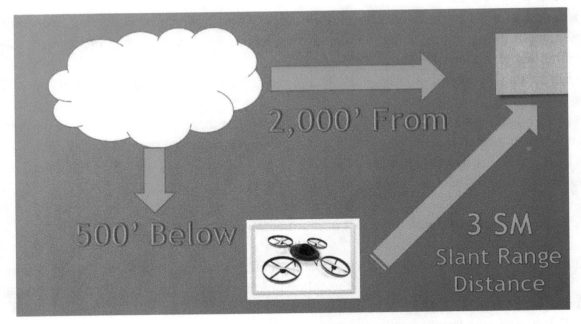

Visibility and Cloud Clearances required via CFR 107.51

sUAS aircraft have their own visibility and cloud clearance requirements like manned aircraft. The good news is that it is just one set of requirements, unlike manned aircraft. CFR 107.51 states that sUAS are required to have 3 SM visibility. Specifically the FAA defines it as the, "average slant distance from the control station at which prominent unlighted objects may be seen and identified by day and prominent light objects may be seen at night (waiver).

Here is an example of visibility and cloud clearance requirements: You arrive on the flight mission site and there are clouds at 700' AGL. How high can you legally fly? Answer: 200' AGL (700' – 500' = 200' AGL).

Subpart C—Remote Pilot Certification

§ 107.53 Applicability.

This subpart prescribes the requirements for issuing a remote pilot certificate with a small UAS rating.

§ 107.57 Offenses involving alcohol or drugs.

(a) A conviction for the violation of any Federal or State statute relating to the growing, processing, manufacture, sale, disposition, possession, transportation, or importation of narcotic drugs, marijuana, or depressant or stimulant drugs or substances is grounds for:

(1) Denial of an application for a remote pilot certificate with a small UAS rating for a period of up to 1 year after the date of final conviction; or

(2) Suspension or revocation of a remote pilot certificate with a small UAS rating.

(b) Committing an act prohibited by § 91.17 (a) or § 91.19 (a) of this chapter is grounds for:

(1) Denial of an application for a remote pilot certificate with a small UAS rating for a period of up to 1 year after the date of that act; or

(2) Suspension or revocation of a remote pilot certificate with a small UAS rating.

§ 107.59 Refusal to submit to an alcohol test or to furnish test results.

A refusal to submit to a test to indicate the percentage by weight of alcohol in the blood, when requested by a law enforcement officer in accordance with § 91.17(c) of this chapter, or a refusal to furnish or authorize the release of the test results requested by the Administrator in accordance with § 91.17(c) or (d) of this chapter, is grounds for:

(a) Denial of an application for a remote pilot certificate with a small UAS rating for a period of up to 1 year after the date of that refusal; or

(b) Suspension or revocation of a remote pilot certificate with a small UAS rating.

§ 107.61 Eligibility.

Subject to the provisions of §107. 57 and § 107.59, in order to be eligible for a remote pilot certificate with a small UAS rating under this subpart, a person must:

(a) Be at least 16 years of age;

(b) Be able to read, speak, write, and understand the English language. If the applicant is unable to meet one of these requirements due to medical reasons, the FAA may place such operating limitations on that applicant's certificate as are necessary for the safe operation of the small unmanned aircraft;

(c) Not know or have reason to know that he or she has a physical or mental condition that would interfere with the safe operation of a small unmanned aircraft system; and

(d) Demonstrate aeronautical knowledge by satisfying one of the following conditions:

(1) Pass an initial aeronautical knowledge test covering the areas of knowledge specified in §107. 73 (a); or

(2) If a person holds a pilot certificate (other than a student pilot certificate) issued under part 61 of this chapter and meets the flight review requirements specified in § 61.56, complete an initial training course covering the areas of knowledge specified in § 107.74 (a) in a manner acceptable to the Administrator.

§ 107.63 Issuance of a remote pilot certificate with a small UAS rating.

An applicant for a remote pilot certificate with a small UAS rating under this subpart must make the application in a form and manner acceptable to the Administrator.

(a) The application must include either:

(1) Evidence showing that the applicant passed an initial aeronautical knowledge test. If applying using a paper application, this evidence must be an airman knowledge test report showing the passage of the knowledge test; or

(2) If a person holds a pilot certificate (other than a student pilot certificate) issued under part 61 of this chapter and meets the flight review requirements specified in § 61.56, a certificate of completion of a part 107 initial training course.

(b) If the application is being made pursuant to paragraph (a) (2) of this section:

(1) The application must be submitted to a Flight Standards District Office, a designated pilot examiner, an airman certification representative for a pilot school, a certificated flight instructor, or other person authorized by the Administrator;

(2) The person accepting the application submission must verify the identity of the applicant in a manner acceptable to the Administrator; and

(3) The person making the application must, by logbook endorsement or other manner acceptable to the Administrator, show the applicant meets the flight review requirements specified in § 61.56 of this chapter.

§ 107.64 Temporary Certificate

(a) A temporary remote pilot certificate with a small UAS rating is issued for up to 120 calendar days, at which time a permanent certificate will be issued to a person whom the Administrator finds qualified under this part

(b) A temporary remote pilot certificate with a small UAS rating expires:

(1) On the expiration date shown on the certificate;

(2) Upon receipt of the permanent certificate; or

(3) Upon receipt of a notice that the certificate sought is denied or revoked.

§ 107.65 Aeronautical knowledge recency.

A person may not operate a small unmanned aircraft system unless that person has completed one of the following, within the previous 24 calendar months:

(a) Passed an initial aeronautical knowledge test covering the areas of knowledge specified in § 107.73 (a);

(b) Passed a recurrent aeronautical knowledge test covering the areas of knowledge specified in § 107.73(b); or

(c) If a person holds a pilot certificate (other than a student pilot certificate) issued under part 61 of this chapter and meets the flight review requirements specified in § § 61.56, passed either an initial or recurrent training course covering the areas of knowledge specified in § 107.74 (a) or (b) in a manner acceptable to the Administrator.

§ 107.67 Knowledge tests: General procedures and passing grades.

(a) Knowledge tests prescribed by or under this part are given by persons and in the manner designated by the Administrator.

(b) An applicant for a knowledge test must have proper identification at the time of application that contains the applicant's:

(1) Photograph;

(2) Signature;

(3) Date of birth, which shows the applicant meets or will meet the age requirements of this part for the certificate and rating sought before the expiration date of the airman knowledge test report; and

(4) Permanent mailing address. If the applicant's permanent mailing address is a post office box number, then the applicant must also provide a current residential address.

(c) The minimum passing grade for the knowledge test will be specified by the Administrator.

§ 107.69 Knowledge tests: Cheating or other unauthorized conduct.

(a) An applicant for a knowledge test may not:

(1) Copy or intentionally remove any knowledge test;

(2) Give to another applicant or receive from another applicant any part or copy of a knowledge test;

(3) Give or receive assistance on a knowledge test during the period that test is being given;

(4) Take any part of a knowledge test on behalf of another person;

(5) Be represented by, or represent, another person for a knowledge test;

(6) Use any material or aid during the period that the test is being given, unless specifically authorized to do so by the Administrator; and

(7) Intentionally cause, assist, or participate in any act prohibited by this paragraph.

(b) An applicant who the Administrator finds has committed an act prohibited by paragraph (a) of this section is prohibited, for 1 year after the date of committing that act, from:

(1) Applying for any certificate, rating, or authorization issued under this chapter; and

(2) Applying for and taking any test under this chapter.

(c) Any certificate or rating held by an applicant may be suspended or revoked if the Administrator finds that person has committed an act prohibited by paragraph (a) of this section.

§ 107.71 Retesting after failure.

An applicant for a knowledge test who fails that test may not reapply for the test for 14 calendar days after failing the test.

§ 107.73 Initial and recurrent knowledge test.

(a) An initial aeronautical knowledge test covers the following areas of knowledge:

(1) Applicable regulations relating to small unmanned aircraft system rating privileges, limitations, and flight operation;

(2) Airspace classification, operating requirements, and flight restrictions affecting small unmanned aircraft operation;

(3) Aviation weather sources and effects of weather on small unmanned aircraft performance;

(4) Small unmanned aircraft loading;

(5) Emergency procedures;

(6) Crew resource management;

(7) Radio communication procedures;

(8) Determining the performance of small unmanned aircraft;

(9) Physiological effects of drugs and alcohol;

(10) Aeronautical decision-making and judgment;

(11) Airport operations; and

(12) Maintenance and preflight inspection procedures.

(b) A recurrent aeronautical knowledge test covers the following areas of knowledge:

(1) Applicable regulations relating to small unmanned aircraft system rating privileges, limitations, and flight operation;

(2) Airspace classification and operating requirements and flight restrictions affecting small unmanned aircraft operation;

(3) Emergency procedures;

(4) Crew resource management;

(5) Aeronautical decision-making and judgment;

(6) Airport operations; and

(7) Maintenance and preflight inspection procedures.

§ 107.74 Initial and recurrent training courses.

(a) An initial training course covers the following areas of knowledge:

(1) Applicable regulations relating to small unmanned aircraft system rating privileges, limitations, and flight operation;

(2) Effects of weather on small unmanned aircraft performance;

(3) Small unmanned aircraft loading;

(4) Emergency procedures;

(5) Crew resource management;

(6) Determining the performance of small unmanned aircraft; and

(7) Maintenance and preflight inspection procedures.

(b) A recurrent training course covers the following areas of knowledge:

(1) Applicable regulations relating to small unmanned aircraft system rating privileges, limitations, and flight operation;

(2) Emergency procedures;

(3) Crew resource management; and;

(4) Maintenance and preflight inspection procedures.

§ 107.77 Change of name or address.

(a) Change of Name. An application to change the name on a certificate issued under this subpart must be accompanied by the applicant's:

(1) Remote pilot certificate with small UAS rating; and

(2) A copy of the marriage license, court order, or other document verifying the name change.

(b) The documents in paragraph (a) of this section will be returned to the applicant after inspection.

(c) Change of address. The holder of a remote pilot certificate with small UAS rating issued under this subpart who has made a change in permanent mailing address may not, after 30 days from that date, exercise the privileges of the certificate unless the holder has notified the FAA of the change in address using one of the following methods:

(1) By letter to the FAA Airman Certification Branch, P.O. Box 25082, Oklahoma City, OK 73125 providing the new permanent mailing address, or if the permanent mailing address includes a post office box number, then the holder's current residential address; or

(2) By using the FAA website portal at www.faa.gov providing the new permanent mailing address, or if the permanent mailing address includes a post office box number, then the holder's current residential address.

§ 107.79 Voluntary surrender of certificate.

(a) The holder of a certificate issued under this subpart may voluntarily surrender it for cancellation.

(b) Any request made under paragraph (a) of this section must include the following signed statement or its equivalent: "I voluntarily surrender my remote pilot certificate with a small UAS rating for cancellation. This request is made for my own reasons, with full knowledge that my certificate will not be reissued to me unless I again complete the requirements specified in §§ 107.61 and 107.63."

Subpart D – Waivers

§ 107.200 Waiver policy and requirements.

(a) The Administrator may issue a certificate of waiver authorizing a deviation from any regulation specified in § 107.205 of this subpart if the Administrator finds that a proposed small UAS operation can safely be conducted under the terms of that certificate of waiver.

(b) A request for a certificate of waiver must contain a complete description of the proposed operation and justification that establishes that the operation can safely be conducted under the terms of a certificate of waiver.

(c) The Administrator may prescribe additional limitations that the Administrator considers necessary.

(d) A person who receives a certificate of waiver issued under this section:

(1) May deviate from the regulations of this part to the extent specified in the certificate of waiver; and

(2) Must comply with any conditions or limitations that are specified in the certificate of waiver.

§ 107.205 List of regulations subject to waiver.

A certificate of waiver issued pursuant to § 107.200 of this subpart may authorize a deviation from the following regulations of this part:

Sec.621

107.25 – Operation from a moving vehicle or aircraft. However, no waiver of this provision will be issued to allow the carriage of property of another by aircraft for compensation or hire.

107.29 – Daylight operation.

107.31 – Visual line of sight aircraft operation. However, no waiver of this provision will be issued to allow the carriage of property of another by aircraft for compensation or hire.

107.33 – Visual observer.

107.35 – Operation of multiple small unmanned aircraft systems.

107.37 (a) – Yielding the right of way.

107.39 – Operation over people.

107.41 – Operation in certain airspace.

107.51 – Operating limitations for small unmanned aircraft.

Regulations, like airspace, are one of the areas that require consistent study. Feel free to use the members' forum at; www.RemotePilotAssociation.com, to get more information about regulations.

Chapter 11

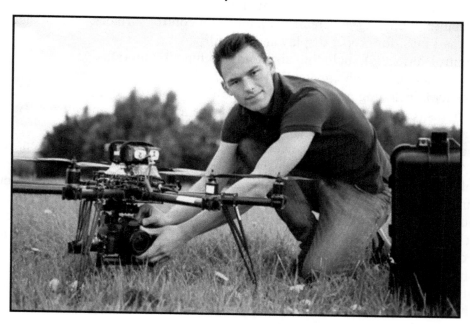

Pre-flight and Maintenance of Your Drone

Manned aircraft have specific maintenance requirements that have to be accomplished by FAA certified Aircraft Maintenance Technicians (AMTs). Fortunately, the approach taken with sUAS is much more simplified but carries some of the same themes such as the requirement for; manufacturer prescribed maintenance, software upgrades/updates, replacement, overhaul and repair procedures, time-limited parts, inspection and documentation. The FAA has deemed the manufacturer as the main driver with regards to the care of a UA. Another parallel to manned aircraft maintenance is the necessity for proper documentation of maintenance. Before we get ahead of ourselves, let us talk about what all flights begin with, a pre-flight (safety) inspection.

Pre-Flight (safety) inspections are what some manned aircraft pilots call a "walk around." The function of a pre-flight inspection is to insure the UA is **airworthy** (Memory aid: Worthy of the Air) meaning in a **condition safe for flight**. It's not fun to get airborne and find out something is wrong with your aircraft. To illustrate, a UA, Remote Pilot (RP) wants to quickly get in the air, takes-off and to find one of the propellers on his quadcopter wasn't secured properly. The blade departs the aircraft becomes difficult to control resulting in a crash landing into an expensive automobile causing $1,000 of damage. Since the amount was greater than $500 (and the automobile was worth a lot more to replace) it was required to be reported to the FAA. Post FAA investigation revealed that an adequate pre-flight would have prevented that accident.

A UA pilot should follow the manufacturers' guidance, if provided, as far as the procedures for doing a pre-flight. If manufacturer checklist isn't provided the FAA provides a basic checklist, the items are:

1. Visual condition inspection of the UAS components;
2. Airframe structure (including undercarriage), all flight control surfaces, and linkages;
3. Registration markings, for proper display and legibility;
4. Moveable control surface(s), including airframe attachment point(s);
5. Servo motor(s), including attachment point(s);
6. Propulsion system, including powerplant(s), propeller(s), rotor(s), ducted fan(s), etc.;
7. Verify all systems (e.g., aircraft and control unit) have an adequate energy supply for the intended operation and are functioning properly;
8. Avionics, including control link transceiver, communication/navigation equipment, and antenna(s);
9. Calibrate UAS compass prior to any flight;
10. Control link transceiver, communication/navigation data link transceiver, and antenna(s);
11. Display panel, if used, is functioning properly;
12. Check ground support equipment, including takeoff and landing systems, for proper operation;
13. Check that *the* control link correct functionality is established between the aircraft and the CS;
14. Check for correct movement of control surfaces using the CS;
15. Check on-board navigation and communication data links;
16. Check flight termination system, if installed;
17. Check fuel for correct type and quantity;
18. Check battery levels of the aircraft and CS;
19. Check that any equipment, such as a camera, is securely attached;
20. Verify communication with UAS and that the UAS has acquired a GPS location from at least four satellites;
21. Start the UAS propellers to inspect for any imbalance or irregular operation;
22. Verify all controller operation for heading and altitude;
23. If required by flight path walk through, verify any noted obstructions that may interfere with the UAS; and
24. At a controlled low altitude, fly within range of any interference and recheck all controls and stability.

Tip: Use the manufacturer's checklist (or FAA checklist in the absence of a manufacturer provided one) utilizing the team communication concept discussed in the Communications chapter called Challenge and Response. The Person Manipulating the Controls (PMC) reads an item (in-order) out-loud (Challenge) to the Remote Pilot (RP) that is from the aircraft in a position to check the various components of the aircraft. The RP Responds to the PMC Challenge by stating what has been checked or verified. For example, The PMC states, "landing gear skids (Challenge)," the RP response, "secure" (Response).

There are **2 Types of Maintenance**:

*Scheduled Maintenance is maintenance, or preventative maintenance, that the manufacturer has set forth in a schedule. Very similar to what you would find in the owner's manual of an automobile.

*Unscheduled Maintenance is the time the aircraft experiences a mechanical or software issue during the course of regular flight operations (or caught during pre-flight checks) in-between regular scheduled maintenance.

Note: In the absence of schedule and unscheduled maintenance protocol from the manufacturer, use industry best practices and guidance.

Replacement of Parts with Alternative Parts is permitted with one, according to the FAA, "that is in a condition for safe operation." If this is not possible the UA should be replaced.

Who Can Perform Maintenance?

Maintenance to a UA can be performed by competent personnel be it the owner, manufacturer or other qualified personnel. The FAA suggests seeking out maintenance personnel that have professional certifications, even going as far as suggesting FAA certificated Airframe and Powerplant (A&P) Mechanics, repairmen, repair stations or those working under A&P mechanics and repairmen. While I doubt the validity of pulling up to the local manned aircraft shop to find anyone that is familiar with sUAS. The only exception would be an Avionics Repair shop that deals with hardware and software of electrical/electronic components. While still very different some the general skill sets are complementary. I would rather seek out someone that specializes in consumer/prosumer aircraft, like a competent hobby shop.

Documentation of Maintenance

This is the area that can get someone on the wrong side of FAA regulations. In the event of an accident the feds (that's what I affectionately call them) will more than likely want to review your maintenance records. Manned aircraft have two main maintenance logbooks (logs), Airframe and Powerplant. sUAS aircraft should be able to get away with just one maintenance log book. According to the FAA, maintenance documentation includes, "any repair, modification, overhaul, or replacement of a system component resulting from normal flight operations, and recording the time-in-service for that component at the time of the maintenance procedures."

The FAA gives requirement on how to keep maintenance records by stating, "Recordkeeping that includes a record of all periodic inspections, maintenance, preventative maintenance, repairs, and alterations performed on the sUAS could be retrievable from either hard copy and/or electronic logbook format for future reference. This includes all components of the sUAS, including: small UA, *Control Station* (CS), launch and recovery equipment, C2 link equipment, payload, and any other components required to safely operate the sUAS."

The good news is that by keeping records you can track how long components will last over time, for example, electric propeller motors. This tracking called "time-in-service," typically logged in hours. An alternative, or addition way, is to track cycling. For example, one takeoff and one landing is a complete cycle.

What is the best way to do maintenance recordkeeping? It can be as easy as creating an Excel file, commercial specialized software (cloud-based or otherwise) or an old school physical aircraft maintenance log. Any of the methods will include; date, describing the maintenance performed, total flight hours, total component hours (if applicable) and whom the work was performed by (w/signature).

A Word About Lithium Batteries and Battery Fires

Lithium-based batteries are highly flammable and capable of ignition. A battery fire could cause an in-flight emergency by causing a Loss of Control (LOC) of the small UA. Lithium battery fires can be caused when a battery short circuit, is improperly charged, is heated to extreme temperatures, is damaged as a result of a crash, is mishandled, or is simply defective. The Remote PIC should consider following the manufacturer's recommendations, when available, to help ensure safe battery handling and usage.

The risk of Lithium batteries doesn't end on the mission site, but also during transportation to the job site, especially when using the airlines or cargo airlines for transport. While most batteries that are installed in a computer, camera and the UA are permitted, the FAA has concerns about the transport of supplemental (extra) battery packs. They have thus developed guidelines about packaging and number of batteries that can be carried in one area. Check with the airline you are flying with and read these FAA documents on the topic:

*Risks in Transporting Lithium Batteries in Cargo by Aircraft,
http://www.faa.gov/other_visit/aviation_industry/airline_operators/airline_safety/safo/all_safos/media/2010/SAFO10017.pdf.

* Carriage of Spare Lithium Batteries in Carry-on and Checked Baggage,
http://www.faa.gov/other_visit/aviation_industry/airline_operators/airline_safety/safo/all_safos/media/2015/SAFO15010.pdf.

*Fighting Fires Caused By Lithium Type Batteries in Portable Electronic Devices,
http://www.faa.gov/other_visit/aviation_industry/airline_operators/airline_safety/safo/all_safos/media/2009/SAFO09013.pdf.

Frequency and Signal Strength of UA

From AC 107-2, sUAS Frequency Utilization states:

An sUAS typically uses radio frequencies (RF) for the communication link between the CS and the small UA.

1) Frequency spectrum (RF) Basics. The 2.4 GHz and 5.8 GHz systems are the unlicensed band RFs that most sUAS use for the connection between the CS and the small UA. Note the frequencies are also used for computer wireless networks and the interference can cause problems when operating a UA in an area (e.g., dense housing and office buildings) that has many wireless signals. Loss of Control (LOC) and flyaways are some of the reported problems with sUAS frequency implications.

To avoid frequency interference, many modern sUAS operates using a 5.8 GHz system to control the small UA and a 2.4 GHz system to transmit video and photos to the ground. Consult the sUAS operating manual and manufacturers recommended procedures before conducting sUAS operations.

It should be noted that both RF bands (2.4 GHz and 5.8 GHz) are considered line of sight and the command and control link between the CS and the small UA will not work properly when barriers are between the CS and the UA. Part 107 requires the remote PIC or person manipulating the controls to be able to see the UA at all times, which should also help prevent obstructions from interfering with the line of sight frequency spectrum.

Checking Airworthiness of Your sUAS

If the manufacturer of your UA doesn't include a maintenance troubleshooting guide the FAA have created a generic one that can be used in a flight operation. Check out the guide below:

Condition	Action
1. Structural or skin cracking	Further inspect to determine scope of damage and existence of possible hidden damage that may compromise structural integrity. Assess the need and extent of repairs that may be needed for continued safe flight operations.
2. Delamination of bonded surfaces	Further inspect to determine scope of damage and existence of possible hidden damage that may compromise structural integrity. Assess the need and extent of repairs that may be needed for continued safe flight operations.
3. Liquid or gel leakage	Further inspect to determine source of the leakage. This condition may pose a risk of fire resulting in extreme heat negatively impacting aircraft structures, aircraft performance characteristics, and flight duration. Assess the need and extent of repairs that may be needed for continued safe flight operations.
4. Strong fuel smell	Further inspect to determine source of the smell. Leakage exiting the aircraft may be present and/or accumulating within a sealed compartment. This condition may pose a risk of fire resulting in extreme heat negatively impacting aircraft structures, aircraft performance characteristics, and flight duration. Assess the need and extent of repairs that may be needed for continued safe flight operations.
5. Smell of electrical burning or arcing	Further inspect to determine source of the possible electrical malfunction. An electrical hazard may pose a risk of fire or extreme heat negatively impacting aircraft structures,

	aircraft performance characteristics, and flight duration. Assess the need and extent of repairs that may be needed for continued safe flight operations.
6. Visual indications of electrical burning or arcing (black soot tracings, sparking)	Further inspect to determine source of the possible electrical malfunction. An electrical hazard may pose a risk of fire or extreme heat negatively impacting aircraft structures, aircraft performance characteristics, and flight duration. Assess the need and extent of repairs that may be needed for continued safe flight operations.
7. Noticeable sound (decibel) change during operation by the propulsion system	Further inspect entire aircraft with emphasis on the propulsion system components (i.e., motors and propellers) for damage and/or diminished performance. Assess the need and extent of repairs that may be needed for continued safe flight operations.
8. Control inputs not synchronized or delayed	Discontinue flight and/or avoid further flight operations until further inspection and testing of the control link between the ground control unit and the aircraft. Ensure accurate control communications are established and reliable prior to further flight to circumvent possible loss of control resulting in the risk of a collision or flyaway. Assess the need and extent of repairs that may be needed for continued safe flight operations.
9. Battery casing distorted (bulging)	Further inspect to determine integrity of the battery as a reliable power source. Distorted battery casings may indicate impending failure resulting in abrupt power loss and/or explosion. An electrical hazard may be present, posing a risk of fire or extreme heat negatively impacting aircraft structures, aircraft performance characteristics, and flight duration. Assess the need and extent of repairs that may be needed for continued safe flight operations.
10. Diminishing flight time capability (electric powered propulsion systems)	Further inspect to determine integrity of the battery as a reliable power source. Diminishing battery capacity may indicate impending failure due to exhausted service life, internal, or external damage. An electrical hazard may
	be present, posing a risk of fire or extreme heat negatively impacting aircraft structures, aircraft performance characteristics, and flight duration. Assess the need and extent of repairs that may be needed for continued safe flight operations.

11. Loose or missing hardware/fasteners	Further inspect to determine structural integrity of the aircraft and/or components with loose or missing hardware/fasteners. Loose or missing hardware/fasteners may pose a risk of negatively impacting flight characteristics, structural failure of the aircraft, dropped objects, loss of the aircraft, and risk to persons and property on the grounds. For continued safe flight operations, secure loose hardware/fasteners. Replace loose hardware/fasteners that cannot be secured. Replace missing hardware/fasteners.

Bulging (or swelling) battery casings can mean the integrity of the unit has been compromised. This could lead to an explosion or rapid power loss, a definite safety risk. Apparently there has been an overheat condition that could affect surrounding parts. Inspect the battery further and determine the best action for repair.

Chapter 12

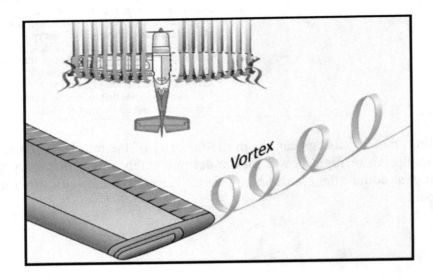

Aerodynamics

Understanding aerodynamics will allow you to understand why the aircraft performs as it does. There is no doubt in my mind that there are some great pilots that came up from the Remote Control (RC) ranks that know a lot about aerodynamics so this will be a review. This value in this section is learning Fed Speak (not an actual term, I just make it up), on what the FAA is looking for in key phrasing, etc. A quick look at the parts of the aircraft then onto the Four Forces of Flight.

Parts of the Aircraft

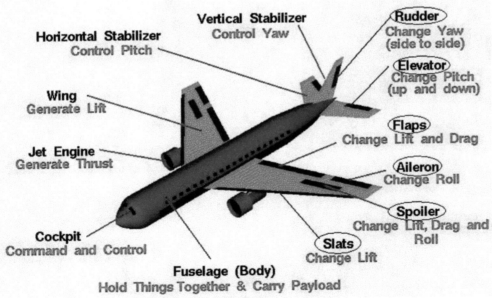

Parts of an Airplane (source: NASA)

If you are sporting jet engines on your UA my hat is off to you, not a great representation, but this graphic hits the main components which will come up in our advanced aerodynamic test questions.

4 Forces of Flight

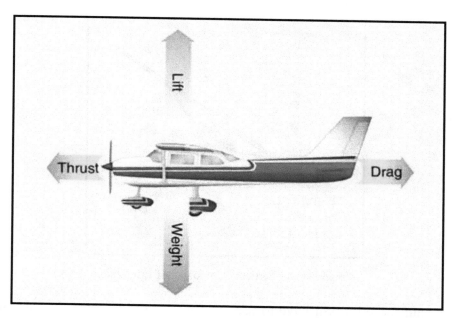

Four Forces of Flight (source: FAA)

The Four Forces of Flight are Lift, Weight (gravity), Thrust and Drag. Lift is diametrically opposed to Weight and Thrust is opposed to Drag. When the aircraft is in straight and level flight all of the forces are in equilibrium. However, things change when aircraft turn, climb and descend.

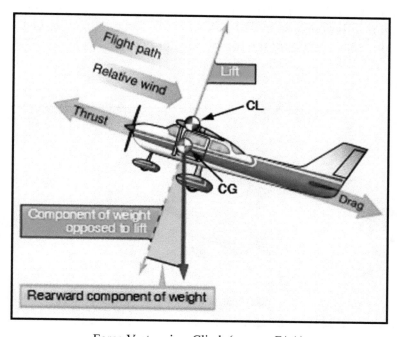

Force Vectors in a Climb (source: FAA)

When an aircraft turns the wings lose lift and the pilot has to use the Elevator control to maintain altitude (if that is what he/she is wanting to do, let us just say it is).

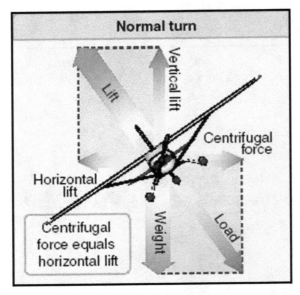

Horizontal and Vertical Component of Lift (source: FAA)

When turning we said the pilot will have to use an elevator to overcome the loss of lift because of increases in centrifugal force and weight. Using the elevator increases the horizontal and vertical component of lift, creating more total lift.

G-force, a.k.a. Load Factor

Turns increase G-force, a.k.a. Load Factor, that push down in the seat, feeling you get when going to the bottom of a hill on a roller coaster, called a Positive G-force. Negative G-force (G's for short) are the push you out of your seat you get at the top of a roller coaster.

AXIS of the Aircraft

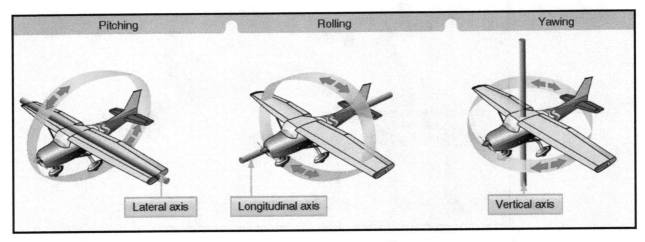

Aircraft Axis (source: FAA)

***Lateral Axis** of the aircraft travels from wingtip to wingtip roughly along the same path as the backbone support of the wing called the spar. If you hold the imaginary rod, if you will, stationary you see it rotates in the Pitch motion which is controlled by the Elevator (or Stabilator if the entire surface moves).

***Longitudinal Axis** run the length of the fuselage (main body) of the aircraft and allows the aircraft to Roll by means of the ailerons. Memory aid: I like to call the longitudinal axis the "armrest" axis to remember the direction of the imaginary line through the fuselage.

***Vertical Axis** of the aircraft proceeds vertically through the fuselage at approximately the Center of Gravity (CG). Movement about the Vertical Axis is called Yaw (no Y'all for you Southerners out there), controlled by the Rudder.

Angle of Attack (AOA)

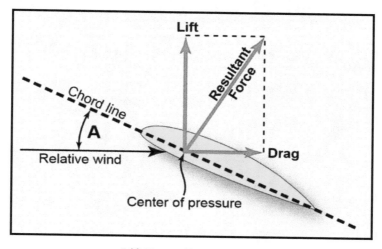

Lift Vector (Source: FAA)

Angle of Attack (AOA) is used as an indicator to illustrate when a wing is nearing stall, called Critical Angle of Attack. A stall condition is when the airflow over the wing becomes disturbed and as a result loses lift. Stalling of a wing typically happens, on average, at 15° of pitch. This number is derived from the Chord Line and the Relative Wind (see Figure 1 above). Chord Line is the imaginary line made by dissecting the airfoil cross-section into two halves. Relative Wind is the direction from which the incoming air is approaching the wing. Memory Aid: The old joke is that Relative Wind isn't your Momma screaming at you. Corny, but you will remember it. See an old NASA wind tunnel video about stalls from the 1930's: https://youtu.be/3_WgkVQWtno.

Center of Pressure (CP) is a centralized location (See the Force Vectors graphic a few illustrations back) over the airfoil in which all of the lift is concentrated, a.k.a. Center of Lift (CL).

Center of Gravity (CG) is the balance point of the aircraft. If you have put together a simple balsawood glider as a kid and tried to balance the fuselage (main body) on your finger, that is Center of Gravity (CG), check out the Force Vectors graphic again. Notice in the graphic that the CL (a.k.a. CP) and the CG are close in proximity to each other. The most stable relationship (when did personal relationships come into the picture) is when the CP is slightly behind the CG. However, CG can change depending on how the aircraft is loaded (or weighted if you will). The FAA states this relationship by saying, "designers fix the aft limit of the CG forward of the CP for the corresponding flight speed in order to retain flight equilibrium." See the "CG and CL (a.k.a. CP) relationship" graphic to see different scenarios.

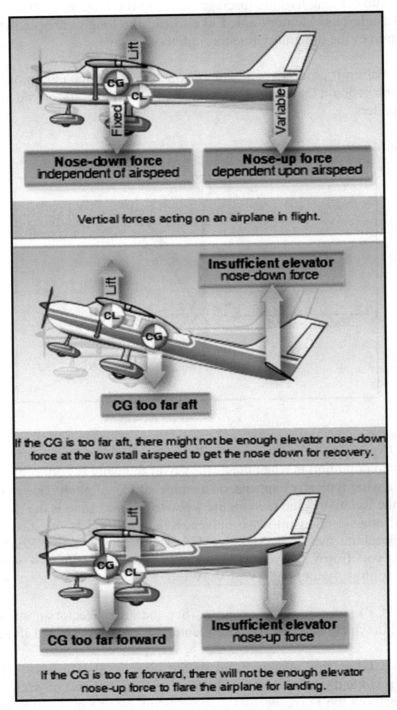

CG and CL (a.k.a. CP) Relationship in various positions (Source FAA)

We want to clarify one concept in the graphic above (CG and CL) and remind folks there is a natural down force on the tail surface that works in a symbiotic relationship (under a perfect scenario) when the CL is behind the CG. Now for a slightly confusing part you will most likely see on the test.

IMPORTANT: During the explanation of the Axis of the Aircraft (Vertical, Longitudinal and Vertical) we asked you to imagine the "rod" axis as being held stationary in-order to understand what control surface corresponded with the rotation of the axis. Shelve this concept for a moment as we talk about stability.

Stability of an aircraft means that when an aircraft is disrupted it will return to its equilibrium within a minimum of time, the less time the better (called Positive Static and Dynamic Stability). When someone says, the aircraft has **Positive Static and Dynamic Stability about the Longitudinal Axis** it means it is stable about the Pitch axis. So basically what we are telling you to do is think exactly opposite of what we told you regarding the various axis. In other words, take the longitudinal axis and move the imaginary "rod" instead of holding it stationary. I know it's a painful concept, reminds me of English class when they stated, "this rule is true unless it is this exhaustive list of exceptions."

Once we have a good, stable flying aircraft then let us see what happens to AOA at various flight speeds.

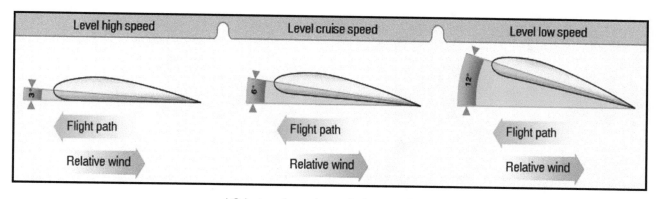

AOA at various airspeeds (source: FAA)

Note that the AOA is least at high speed and greatest at low speeds. If you think about it this makes sense because if there is more air flowing over the wing, there is more lift. More lift means the wing doesn't have to do the heavy lifting that it does at slow speeds, such as on take-off and landing.

Adverse Yaw is the drag that is produced by the down aileron when banking an aircraft. You might be thinking that a down aileron means that the particular wing will create lift, thus raising it up for a bank. However, the polar opposite of lift is drag and any time lifting surfaces create lift, drag is a byproduct. This drag creates a yawing motion and makes for an imperfect, or delayed, turn. To overcome Adverse (meaning bad) Yaw rudder is used to yaw the aircraft in the direction of the desired bank. Remember: Right turn, right rudder – left turn, left rudder.

Flaps are used to increase the camber (surface area) of the wing, working in unison on the inside trailing edge of each wing. The interesting thing about flaps is the ratio of lift to drag they have when deployed to various selections (or degrees). For example, lowering the flaps to 10° and there is more lift than drag created, perfect for making short field take-off's. Now, lower flaps to a full 30° and there is more drag than lift, perfect for making steep approaches to a runway at slower speeds allowing for shorter landing distances over obstacles.

What about Rotorcraft?

We realize that the majority of the readers will be flying a multi-copter of some description (my money is on a quadcopters) but the FAA seems to be defaulting to the fixed wing variety of UA. However, keep in mind that multi-rotors, like propellers, are essentially wings that propel the aircraft vertically. Asymmetrical Thrust is used to maneuver the rotorcraft by adjusting the speed (thrust) of the rotors to turn the aircraft. Take-off's and hover type landings are accomplished through a more even power increase or reduction respectively. Rotorcraft are affected by torque and P-factor like their fixed wing cousins, but it is all worked out by the software and controller. We don't mean to seem like we are playing favorites here, but we wanted to keep the main thing, the test, the main thing.

Remote *Drone* Pilot Study Guide ~ Practice Test Questions

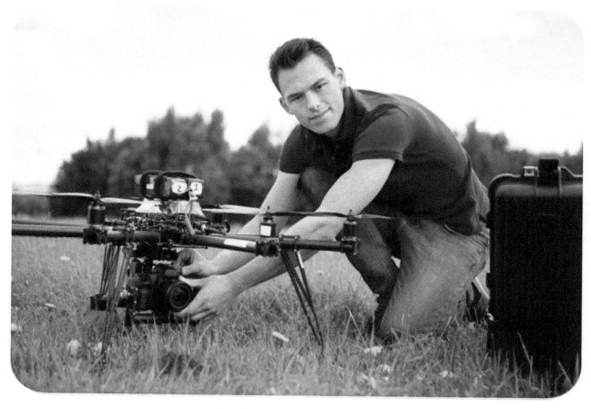

Quadcopter (Source: CanStockPhoto)

Study Tip: Go through the questions slowly in the beginning, study the answers, make notes and develop your memory aids (written and drawn). After going through the questions several times, jot down questions you are having problems answering correctly. Go back to the chapters and study information on the topic(s). Shoot for correctly answering the questions 90% of the time, which allows enough of a buffer to pass the FAA Knowledge test with flying colors.

Setting the Table ~ Introduction
Definitions of sUAS

1. To be defined as a sUAS it must weight under:

a) 4.4 lbs.

b) 57 lbs.

c) 55 lbs.

Answer – c. A small Unmanned Aerial System (sUAS) is required to weigh less than 55 lbs. This weight includes all external attachments and accessories. Reference: CFR 107.3

Excluded Aircraft and Operations

2. Which of the following are excluded from sUAS Part 107?

a) Model R/C aircraft

b) Amateur rockets

c) All of the above

Answer – c. Model R/C, amateur rockets, moored and unmanned free balloons, kites, operations outside of the US, public aircraft operations and air carrier operations are excluded from Part 107 regulations.

Commercial or Hobby Flight

3. Flying for fun with no compensation is covered under 14 CFR Part 107?

a) True

b) False

Answer – b. Flying for fun, or as a hobby, is not considered commercial flying and not covered in 14 CFR Part 107.

4. A drone pilot is asked by a concert promoter to take video footage of a large music event they are hosting in exchange for 3 All-Access VIP passes for the weeklong event. This is considered commercial flying?

a) True

b) False

Answer – a. The exchange of the passes in lieu of cash or check is still considered payment which constitutes commercial flight. Furthermore, the use of the video is for the furtherance of a business interest and while not the drone pilot business is considered a commercial flight activity under 14 CFR 107.

5. An agent in the real estate business has a high listing price home; he would like aerial video footage that will be used for a dynamic online video ad. He asks a friend with a drone to handle the project, but the drone pilot says he can't charge for the work. The drone pilot explains that the FAA might issue a fine if they catch him flying a drone commercially. The real estate representative jumps at the chance for free aerial video work and the drone pilot is happy to do him a favor. This is **not** considered as flying commercial drone.

a) True

b) False

Answer – b. Although the drone pilot is refusing payment the ultimate use of the video will be used in the furtherance of business, in this case, the real estate business owner. This flight will fall under 14 CFR 107 regulations.

6. The FAA is ultimately responsible for federal oversight of the National Airspace System (NAS). However, the "FAA encourages sUAS operators to review

a) *State* laws before operating their UAS."

b) regulations of the sUAS aircraft country of manufacture."

c) Standard Operating Procedures (SOP) developed by the sUAS Standards Board (sUAS-SB)."

Answer – a. Many States created laws for drones (sUAS) before the FAA to handle specific issues such as; safety, security, privacy, property protection, etc. After the FAA regulations had come into effect (and became clarified) States were encouraged not to overlap federal law in the process of creating laws that they had jurisdiction. As a result, the FAA encourages drone operators also to review State law. Note: This review should extend to municipalities where appropriate. Reference: AC 107-2 (1.1.3, 5.13.3)

7. The FAA gives guidance that sUAS operators are to review and be aware of State laws that govern sUAS. While individual States can have their own unique laws, most address

a) only the applicable fees associated with commercial sUAS flight operations.

b) How commercial and hobbyist sUAS flight operations can defer to the FAA website for regulatory requirements.

c) State level testing certification for sUAS operators that covers protection of; person's privacy, wildlife, airspace, and property. Additionally, States will typically define the distinction between commercial and hobbyist flight operations.

Answer – c. Due to delays in FAA regulation, there is some overlap between federal and State laws regarding sUAS. The FAA has encouraged States only to create legislation in which they have jurisdiction and encourages UA operators to review such laws. Many States laws address common themes of protecting persons, property, wildlife, property and airspace. Reference: AC 107-2 (5.13.3, 1.1.3)

8. Which of the following unmanned aircraft flights would not be considered recreational operation:

a) A corporate researcher is flying a 75-pound, custom-built UAS.

b) A student offering to fly a quadcopter in his neighborhood for home gutter inspection for a fee.

c) A realtor is flying a quadcopter over a new home development to take pictures of their listings for use in advertising.

d) All of the above

Answer – d. Money does not have to change hands to be considered a commercial operation. The furtherance of business, as with these three examples. Reference: AC 107-2

9. State law stipulates that all UAS activities conducted in the National Airspace System (NAS) are governed by federal regulations on aviation activities. Individual States govern only UAS activities over which it has legal authority.

a) TRUE

b) FALSE

Answer – a. The FAA has encouraged States to limit the creation of laws that overlap federal law. However, the FAA does suggest UA operators to review State law before flight. Reference: AC 107-2 (1.1.3)

10. Both the federal government and individual States have jurisdiction concerning crimes committed in which a UAS was used.

a) True

b) False

Answer – a. Crimes committed with a UA can be in violation of State and Federal law. Reference: CFR 107.57, 107.23, 107.19, AC 107-2 (1.1.3)

11. State Law example: What constitutes the weaponization of a UAS?

a) Drones are for fun, so no matter what you do, they won't be considered weaponized.

b) Attaching a weapon to a UAS.

c) Using a weapon from the ground to interfere with a UAS.

Answer – b. This is an obvious question to drive the point that putting any sort of weapon on a UA is illegal in the State and federal level. Reference: State law example, CFR 107.23

12. While FAA Advisory Circulars (ACs), such as Part 107 covering sUAS, do not provide "legal interpretations" of regulations it does provide guidance in the areas of

a) Filling out forms required by State law.

b) Airmen (Remote Pilot) certification, aircraft registration and marking, airworthiness, and the operation of sUAS aircraft in the National Airspace System (NAS) in compliance with the Code of Federal Regulations (CFR), such as Title 14 CFR Part 107.

c) Searching out and finding the local attorney representation specific to aviation in the event of action taken by persons (private and governmental).

Answer – b. Reading Advisory Circulars (ACs) is a good way to better understand, commonly misunderstood regulations. While not considered a legal interpretation, it is a great tool for the lay person. Reference: AC 107-2 (1.1)

13. FAA Advisory Circulars (ACs) are not intended to cover every provision of part 107 but rather intended to

a) provide guidance on those provisions of part 107 where additional information is helpful.

b) be used by FAA Operations Inspectors as a tool for enforcement.

c) be used as an Executive Summary for commercial sUAS corporate management officials.

Answer – a. AC's are not considered legal interpretations, but a good clarification of regulations that are commonly misunderstood. Reference: AC 107-2 (1.1.2)

14. In developing CFR Part 107 the FAA addresses aviation safety about UAS by addressing three key areas:

a) cost-benefit analysis of the UAS industry, Parts Manufacturer Approval (PMA) and Type Certification (TC).

b) regulatory fees, hobbyist pilots and air carrier mitigation.

c) personnel, equipment, and operations.

Answer – c. At early conferences regarding the integrating of UAS into the NAS, the military gave great insight into the logical areas such as personnel, equipment, and operations due to their long history with UA applications in the military. This experience was a good groundwork for civilian application. Reference: AC 107-2 (3.2)

15. Model Aircraft used for hobby, or recreational purposes are listed under Part 101 Subpart E (Model Aircraft). Which of the following rules does hobbyist and sUAS commercial operators (under Part 107) have in common (no waivers)?

a) Aircraft weight must be less than 55 pounds (under general terms), does not interfere with, and gives way to manned aircraft, authorization required to fly into airport controlled airspace via FAA approval, and aircraft is flown within Visual Line of Sight (VLOS).

b) Aircraft weight cannot exceed 55 pounds (under general terms), does not interfere with and gives way to manned aircraft, authorization to fly in airport controlled airspace via Air Traffic Control Tower (ATCT) officials, and aircraft is solely flown by First Person View (FPV).

c) No commonalities exist due to the advanced technology of sUAS verses Model Aircraft, therefore, it's operational requirements are starkly different.

Answer – a. One of the goals of regulation is to differentiate between commercial sUAS and hobbyist type flying. The common themes revolve around the safety and protection of; persons, property and certain types of airspace. The regulations define UA by weight since a larger aircraft has more damage potential and requires broader regulation. UAs are always required to give way to manned aircraft which is different from manned aircraft in which less maneuverable aircraft have the right-of-way over more maneuverable aircraft. As the complexity of airspace in which a UA aircraft operates, for example, Class B, C, and D airspace, the requirement for gaining authorization from the FAA is mandatory. Visual Line of Sight (VLOS) is also a requirement at this time requiring positive control and monitoring of UA aircraft. Reference: AC 107-2 (4.1)

16. Part 107 does not apply to the following:

a) Part 107 applies to all aircraft operating in the National Airspace System (NAS) under the regulatory guidance of the National Transportation Safety Board (NTSB).

b) Model aircraft, amateur rockets, moored balloons, unmanned free balloons, kites, Public aircraft operations, air carrier operations, operations outside of the United States.

c) Model aircraft, amateur rockets, moored balloons, unmanned free balloons, kites, Public aircraft operations, air carrier operations, operations within the United States.

Answer – b. Membership RC hobbyist organizations have been successful at protecting their membership from the more restrictive commercial sUAS operations. The FAA also differentiates those aircraft and operations in which it seeks to clarify doesn't fall under Part 107 rules. Reference: AC 107-2 (4.1)

17. For commercial sUAS flight operations, what is the definition of a Control Station (CS)?

a) An interface used by the Remote Pilot (RP) and/or the Person Manipulating the Controls (PMC) to control the flight path of the sUAS.

b) The home office address of the commercial sUAS flight operations as required for FAA Flight Standards District Office (FSDO) auditing process.

c) Control Station (CS) is the official phraseology used by the Visual Observer (VO) when a hazard has been identified, manned aircraft or physical structure, to positively communicate to the Remote Pilot (RP) and Person Manipulating the Controls (PMC).

Answer – a. A CS can be a hand controller the computer station from which total autonomous operations are controlled. The regulation stipulates the RP should be able to take control via the CS if the need arises. Reference: AC 107-2 (4.2.1)

18. To be classified under Part 107, a sUAS is required to be flown

a) every 14 days to maintain its airworthiness certification.

b) by the Visual Observer (VO) under the direction of the Person Manipulating the Controls (PMC).

c) without the possibility of direct human intervention from within or on the aircraft.

Answer – c. In this question, the FAA is making certain UA pilots are aware of what is, and isn't, considered a UA aircraft. In this case, any control within or on the aircraft is not considered to be a UA. Reference: AC 107-2 (4.2.6)

19. Which preflight item(s) are required to be checked before flight, to ensure the safe and efficient operation of the small Unmanned Aircraft (UA) in the National Airspace System (NAS)?

a) Test of the communication links per manufacturer's instructions.

b) Consistent reporting to the FAA Flight Standards District Office (FSDO) Operations and Airworthiness Inspector regarding pre-authorization and debriefing of each commercial flight per CFR Part 107.

c) Monthly post flight checks of the case used to transport the case of anomalies which can be a direct indicator of airworthiness of the Unmanned Aircraft (UA). If anomalies are found a Materials and Defect Report (MDR) should be submitted via the Airmen Safety Reporting System (ASRS).

Answer – a. While other items are required to be checked, a check of the communications link control is the best answer. The manufacturer will dictate instructions on how to accomplish this important pre-flight task. Reference: AC 107-2 (4.2.7)

20. Who can be subject to FAA actions, including civil sanctions and the suspension or revocation of a certificate of waiver if found to be fraudulent or knowingly providing false records or reports, or otherwise reproduces or alters any records or other information for fraudulent purposes?

a) The Remote Pilot only because he/she is ultimately responsible for sUAS flight operations and holds a FAA certificate.

b) The owner of the commercial sUAS company can have his/her repair certificate suspended or revoked.

c) The landowner over which the last mission in which the infraction was committed. The current law states that the landowner takes legal responsibility and possession of sUAS aircraft when hired for a flight mission and therefore responsible.

Answer – a. The point of this question is to express the importance of keeping good records in which the FAA can review in the case of an audit, specifically by the Remote Pilot as the only FAA certificate holder in the operation. Typically, records will be reviewed by the FAA in the case of an accident. Any efforts to falsify reports can fall under a federal offense, a serious matter needless to say. Reference: AC 107-2 (4.4)

Flight Crew Roles, Certification and Drone (sUAS) Registration

Registration

21. A sUAS does not have to be registered if it weighs less than:

a) 0.55 lbs. or 250 grams

b) 55 lbs. or 25000 grams

c) 4.4 lbs. or 2250 grams

Answer – a. A sUAS does not have to be registered with the FAA if it weighs less than 0.55 lbs. (250 grams). All sUAS that weigh more than this and up to below 55 lbs. has to be registered with the FAA. Remember: Registration of aircraft and certification to fly commercial sUAS, are two separate issues. For example, you own a very small quadcopter that weighs just 200 grams (including a 720P camera) and flies it as part of an aerial photography business. While the aircraft doesn't have to be registered the operator is required to be a certificated Remote Pilot (RP). Reference14 CFR Part 47.

22. The minimum age permitted to register a sUAS is?

a) 12

b) 13

c) 16

Answer – b. A person less than 13 years old cannot register a sUAS with the FAA. Someone meeting the minimum age (13 years old or older) requirement can be allowed to register the sUAS. Remember: The registration of the sUAS and its certification (Remote Pilot) are two separate issues. A person is required to be at least 16 years old to operate as a commercial Remote Pilot (a.k.a. exercise the privileges of a Remote Pilot). However, a person can take the Remote Pilot Knowledge Test (RP KT) at 14 years old. Memory aid: 13 to register a sUAS, 14 to take the RP KT, and 16 to exercise the privileges of a commercial RP. References: 14 CFR Part 48, http://www.faa.gov/uas/media/Part_107_Summary.pdf.

23. Before flight operations are allowed, all of the following pertaining to FAA aircraft registration is required except:

a) A unique identifier issued by the FAA that will be placed on the aircraft or in an easily accessible area accessed without hand tools.

b) The FAA issued registration number is placed on the aircraft in a legible and durable fashion; consisting of engraving, permanent marker or self-adhesive label.

c) Flight Plan; the pilot is required to file a flight plan with the FAA before flight operations.

Answer – c. A flight plan, as used during some manned aircraft flights, is not a required part of the registration or flight. A sUAS registration number is required to be visible or accessible and may be in a compartment that is accessible without hand tools and be legible, durable using engraving, permanent marker or self-adhesive tool. Reference 14 CFR Part 48.

24. If a sUAS is registered in a foreign country and operated by a non-US citizen, are allowed to fly in the US only if:

a) They file a flight plan indicating they are foreign registered aircraft before each flight in the US.

b) Apply to the FAA for a Foreign Aircraft Permit before beginning flight operations in the US.

c) Earn a minimum of an FAA Sport Pilot certificate to meet the qualifications for flight in the US.

Answer – b. An FAA Foreign Aircraft Permit is required if an aircraft is; 1) registered in a foreign country and/or 2) owned by a non-US citizen or who is not a permanent resident. Reference – 14 CFR Part 375.41 and 375.43.

25. Under what condition would a small UA not have to be registered before it is operated in the United States?

a) When the aircraft weighs less than 0.55 pounds at takeoff, including all attachments and accessories affixed to the aircraft.

b) When the aircraft has a takeoff weight that is more than 0.55 pounds, but less than 55 pounds, not including fuel and necessary attachments.

c) All small UAS need to be registered regardless of the weight of the aircraft before, during, or after the flight.

Answer – a. A sUAS that weighs less than 0.55 lbs. (or 250 grams) with all attachments and accessories are exempt from FAA sUAS aircraft registration. Remember that registration is a separate task on the certification of the Remote Pilot (RP) for commercial flight operations. An RP is still required for commercial operation even if the aircraft under the threshold weight for registration. References: FAA UAS registration page; https://registermyuas.faa.gov/, Q30UAS SG

26. According to 14 CFR part 48, when must a person register a small UA with the Federal Aviation Administration?

a) All small civilian UAs weighing greater than 0.55 pounds must be registered regardless of its intended use (for hobby or commercial purposes).

b) When the small UA is used for any purpose other than as a model aircraft.

c) All small UAS need to be registered regardless of the weight of the aircraft before, during, or after the first flight.

Answer – a. A sUAS that weighs less than 0.55 lbs. (or 250 grams) with all attachments and accessories is exempt from FAA sUAS aircraft registration. The maximum sUAS weight is under 55 lbs. We think this question leaves open the upper weight that will allow expansion of the general UA regulations for larger weight categories. Remember that registration (must be a minimum of 13 years old) is a separate task with the certification of the Remote Pilot (RP) for commercial flight operations. References: FAA "Getting Started" web page; http://www.faa.gov/uas/getting_started/, Q31UAS SG

27. According to 14 CFR part 48, when would a small UA owner not be permitted to register it?

a) The owner is less than 13 years of age.

b) All persons must register their small UA regardless of age.

c) If the owner does not have a valid United States driver's license.

Answer – a. If the owner is less than 13 years of age, the sUAS must be registered to someone that is 13 years old or older, such as a parent. Age does not disqualify a person from having to register a sUAS. Remember that registration (must be a minimum of 13 years old) is a separate task with the certification of the Remote Pilot in Command (RPIC) for commercial flight operations. References: FAA sUAS registration web page; https://registermyuas.faa.gov/register, Q32UAS SG

Roles

28. The Remote Pilot (RP) can delegate the ultimate responsibility for traffic avoidance, weather, equipment, loading, and aircraft performance calculation/computation to the Person (pilot) Manipulating the Controls (PMC) and Visual Observer (VO).

a) True

b) False

Answer – b. The RP is ultimately responsible for the flight mission, although the RP may delegate certain duties, such as traffic avoidance, weather, equipment, loading, and aircraft performance computations/calculations to the PMC and VO but the ultimate responsibility rest with the RP.

29. When using a small UA in a commercial operation, who is responsible for doing the pre-flight briefing to the flight crew and approved mission site personnel about emergency procedures?

a) The FAA inspector-in-charge.

b) The lead visual observer.

c) The Remote Pilot.

Answer – c. The Remote Pilot (RP) is the lead position in the sUAS flight crew. The RP is over the Person Manipulating the Controls (PMC) and the Visual Observer(s) (VO). Emergency procedures should be briefed by the RPIC. References: CFR 107.49, Q13UAS SG.

30. According to 14 CFR part 107, who is responsible for determining the performance of a small unmanned aircraft?

a) Remote Pilot.

b) Manufacturer.

c) FAA Operations Inspector.

Answer – a. The Remote Pilot (RP) is at the top of the sUAS flight crew hierarchy, followed by the Person Manipulating the Controls (PMC) and the Visual Observer(s) (VO). The RP is responsible for all issues related to the safety of the flight including determining performance. Performance can be affected by added accessories to the airframe or operating flights at high altitudes, called high-density altitude, with reduced performance due to the thinner air. The RP can delegate duties, but the ultimate responsibility is with the RP. References: CFR 107.49, Q28UAS SG

31. After a Remote Pilot (RP) certificate has been earned what is required to keep it current?

a) Nothing. Once the initial Part 107 test has been passed no further actions are necessary. However, a FAA Operations Inspector is required to administer an Oral and Practical Examination on a pre-scheduled periodic basis.

b) 3 takeoffs and landings every 90 calendar days is required and documented in the RP logbook.

c) Through passing another initial or recurrent knowledge test covering items specified in CFR 107.73 (a) (b) or 107.74 (a) or (b) within the previous 24 calendar months.

Answer – c. Manned aircraft pilots have to complete a bi-annual Flight Review (every 24 calendar months) that consist of a combination of ground and flight time with a flight instructor. Remote Pilots (RPs) will also have to complete a review minus the flight portion via a re-currency course (like this one) or retaking of the initial Remote Pilot Knowledge Test (KT). Reference: CFR 107.65, 107.73 (a) (b), 107.74 (a) (b)

32. If a Remote Pilot (RP) changes his/her permanent mailing address and fails to notify the FAA Airmen Certification Branch of the new address, the RP is entitled to exercise the privileges of his/her certificate for a period of

a) 30 days after the date of the move.

b) 60 days after the date of the move.

c) 90 days after the date of the move.

Answer – a. Due to national security concerns, the FAA, under the strong recommendation of the Transportation Security Administration (TSA), has been working to update airmen and aircraft registration records. As a result, the FAA requires notification of a change in permanent mailing address within 30 calendar days or a loss of the exercise of the privileges of the Remote Pilot (RP) certification. In other words, an RP can potentially lose his/her certificate and/or be subject to FAA enforcement action. Reference: CFR 107.77

33. The final authority as to the operation of a UA is the

a) Federal Aviation Administration.

b) Remote Pilot (RP)

c) Aircraft manufacturer.

Answer – b. The RP is at the top of the sUAS flight crew hierarchy and ultimately responsible for the flight. CFR 107.19

Charts, Airspace and You ~ Airspace

34. Figure 21: What airport is located approximately 47° 40 minutes N latitude and 101° 26 (minutes) W longitude?

a) Mercer County Regional Airport

b) Semshenko Airport

c) Garrison Airport

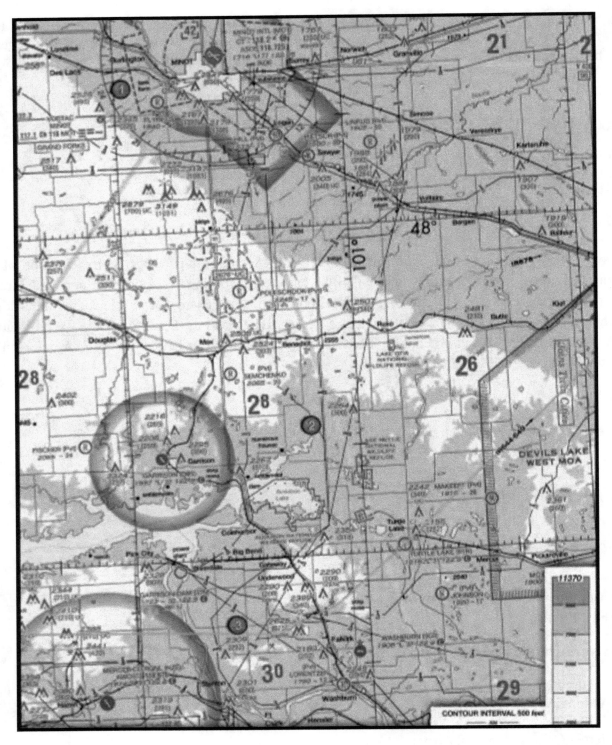

Figure 21. Not to scale.

Answer – c. Horizontal lines of latitude are lower in number traveling toward the equator. Lines of Longitude are vertical lines growing in number traveling from east to west (US example). Look for the reference lat./long. degree number on the chart as a starting point. Remember the line (lat. And long.) not showing a number is the 30-minute reference. References: Aeronautical chart. Q1UAS SGA.

35. Reference Figure 26. What does the line of latitude at Area 4 measures?

a) The degrees of latitude east and west of the Prime Meridian.

b) The degrees of latitude north and south from the equator.

c) The degrees of latitude east and west of the line that passes through Greenwich, England.

Answer – b. The question is more general in nature and doesn't necessarily require a chart reference. Lines of latitude that run horizontally around the earth about the equator. Degree numbers are lower towards the equator and greater towards the poles. References: Aeronautical charts. Q2UAS SG, fig 2. FAA-CT-8080-2G figure 27.

36. Refer to Figure 23, Area 3: What is the floor of the Savannah Class C airspace at the shelf area (outer circle)?

a) 1,300 feet AGL

b) 1,300 feet MSL

c) 1,700 feet MSL.

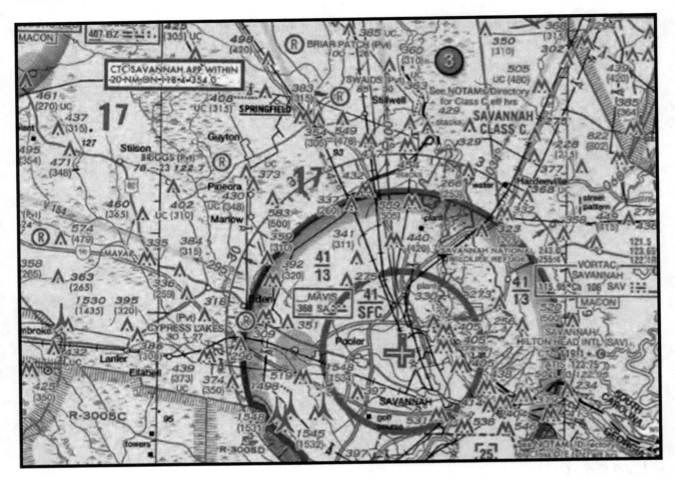

Figure 23. Not to scale.

Answer - a. Look inside of the outer ring of the Savannah Class C ring and you will see the numbers 41/13, which stands for 4100' Mean Sea Level (MSL) and 1300' MSL. 4100' is the top of Class C and 1300' the bottom. Remember to think of an upside down wedding cake to help visualize Class B and C airspace. Altitudes are expressed in MSL since that is what manned aircraft altimeters use in the cockpit as a "working" number or reference. References: Aeronautical Chart, Q3UAS SG, FAA-CT-8080-2G figure 27.

37. Refer to Figure 59, Area 2. The chart shows a gray line with "VR1667, VR1617, VR1638, and VR1668." Could this area present a hazard to the operations of a small UA?
a) No, all operations will be above 400 feet.

b) Yes, this is a Military Training Route from 1,500 feet AGL.

c) Yes, the defined route provides traffic separation to manned aircraft.

144

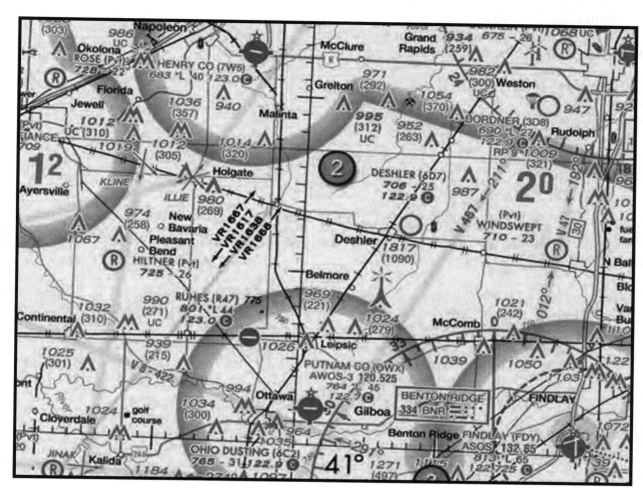

Figure 59, Area 2. Not to scale.

Answer - c. Visual (flight) Routes (VR) are gray lines shown on an aeronautical chart to show where military aircraft training routes are located (think gray camouflage on a fighter jet). VR routes are flown visually by the pilot. If the route has four numbers, as in this example, the route is flown **below** 1500' Above Ground Level (AGL). When the route is three or fewer numbers, it is flown above 1500' AGL. The answer wording is somewhat unclear, but we interpret it to mean that separation (or Mid Air Collision Avoidance – MACA) is mitigated through being listed as a specific route on the chart.

FYI: Instrument (flight) Routes (IR) are shown and operate the same way except that they are flown by the pilot concentrating on the flight instruments and not visually by keeping attention outside of the aircraft. References: Aeronautical Chart index. Q4UAS SG, FAA-CT-8080-2G, http://www.faa.gov/air_traffic/flight_info/aeronav/digital_products/aero_guide/

38. According to 14 CFR part 107 the Remote Pilot of a small unmanned aircraft planning to operate within Class C airspace

a) must use a visual observer

b) is required to file a flight plan

c) is required to receive ATC authorization

Answer – c. Class C airspace has an Air Traffic Control Tower (ATCT) and requires authorization which is done through the FAA website waiver request at https://www.faa.gov/uas/request_waiver/. Reference: CFR 107.41. Q5UAS SG.

39. Refer to Figure 21, East of Area 2. You have been hired by a farmer to use your small UA to inspect his crops. The area that you are to survey is in the Devil`s Lake West MOA, east of area 2. How would you find out if the MOA is active?

a) Refer to the legend for (the) special use airspace phone number.

b) This information is available in the Small UAS database.

c) In the Military Operations Directory.

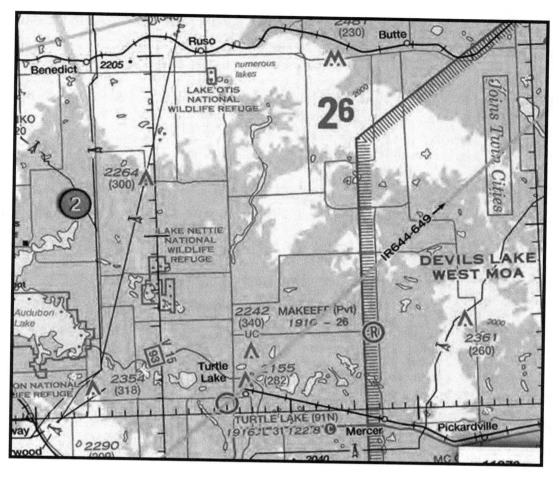

Figure 21, East of Area 2, Devil's Lake West MOA (Not to scale).

Answer - a. The legend of the aeronautical chart will include information on who controls the Devil's Lake West Military Operations Areas (MOA's), hours and altitudes of flight operations. MOA's are large areas of airspace used for military training. The Automated Flight Service Station (AFSS, 1800WXBrief) can also verify if specific MOA's are "hot" (in use) or "cold" (not in use). Reference: Aeronautical Chart. Q6UAS SG.

40. The B4UFLY App for smart-phones are a good way to

a) Get a visual reference regarding the airspace in which a flight mission will take place.

b) To file a flight plan with a direct link to a commercial operator's FAA FSDO Operations Inspector.

c) Review the regulations (CFRs) regarding the Part 107 rule while setting up for a flight mission.

Answer – a. The B4UFLY Application, while not getting good reviews, is a FAA produced tool to keep Remote Pilots (RPs) from flying in airspace that is off limits, requires prior permission or notification in which to operate (such as with FAA prior permission to fly in Class B, C or D airspace).

41. Refer to Figure 20, Area 3: How would a Remote Pilot "CHECK NOTAMS" as noted in the CAUTION box regarding the unmarked balloon?

a) By utilizing the B4UFLY mobile application.

b) By contacting the FAA district office.

c) By obtaining a briefing via an online source such as 1800WXBrief.com.

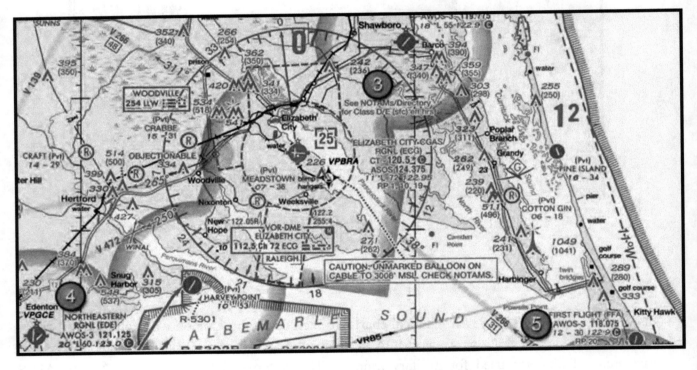

Figure 20, Area 3. Not to scale.

Answer – c. Notices to Airmen (NOTAMS) contain information for flight crews such as altitudes and time of operation for tethered balloons. NOTAMS, are not limited to balloon operations, but also uncharted new towers, runway closures, etc. This information can be requested from the Automated Flight Service Station (AFSS) via phone 1800WXBrief or 1800WXBrief.com. Reference: PHAK, Q7UAS SG.

42. According to 14 CFR part 107, how may a Remote Pilot operate an unmanned aircraft in Class C airspace?

a) The remote pilot must have prior authorization from the Air Traffic Control (ATC) facility having jurisdiction over that airspace.

b) The remote pilot must monitor the Air Traffic Control (ATC) frequency from launch to recovery.

c) The remote pilot must contact the Air Traffic Control (ATC) facility after launching the unmanned aircraft.

Answer – a. CFR 107.41 states that anyone operating in Class C airspace (also Class B, D and within the lateral boundaries of the surface area of Class E airspace designated for an airport) must have prior permission from Air Traffic Control (ATC) via the FAA website waivers form available at; https://www.faa.gov/uas/request_waiver/. Remember, there as many types of airports, (memory aid: Class B, B stands for Big cities, C is for regular Cities and D is for regular airports) that have ATC. Tip to determine if an airport has an ATCT: The appropriate aeronautical chart will show the Control Tower (CT) frequency if it has ATC. The Chart Supplement Guide (formerly the Airport Facility Directory, a.k.a. "the green book") will have detailed information about the airport. References: CFR 107.41, Q33UAS SG

43. A blue segmented circle on a Sectional Chart depicts which class airspace?

a) Class B.

b) Class C.

c) Class D.

Answer – c. The Aeronautical Chart Legend is probably the easiest guide to identify airspace, in this case, Class D. More information can be found in the AIM and CFRs. Class D airspace has a control tower, and it is required to gain permission to fly in this airspace from the FAA waiver request website at; https://www.faa.gov/uas/request_waiver/. Resource: Aero Chart Legend, AIM, PHAK

44. Airspace at an airport with a part-time control tower is classified as Class D airspace only

a) When the weather minimums are below basic VFR.

b) When the associated control tower is in operation.

c) When the associated Flight Service Station is in Operation.

Answer – b. There are many ATC Towers that close at night due to lower traffic levels. When the tower is open, it is considered Class D, the least busy air traffic controlled airspace. Upon the tower closing the airspace is defined as Class E (sometimes G). The aeronautical chart and Airport Supplements are the best resources for more information on the ATCT hours of operations. Resource: AIM, PHAK

45. When a control tower located at an airport within Class D airspace ceases operation for the day, what happens to the airspace designation?

a) The airspace designation normally will not change.

b) The airspace remains Class D airspace as long as a weather observer or automated weather system is available.

c) The airspace reverts to Class E or a combination of Class E and G airspace during the hours the tower is not in operation.

Answer – c. There are many ATC Towers that close at night due to lower traffic levels. When the tower is open, it is considered Class D, the least busy air traffic controlled airspace. Upon the tower closing the airspace is defined as Class E (sometimes G). The aeronautical chart and AFD are the best resources for more information on the ATCT hours of operations. Resource: AIM, PHAK

46. If a control tower and an Automated Flight Service Station (AFSS) are located on the same airport, which function is provided by the AFSS during those periods when the tower is closed?

a) Automatic closing of flight plans.

b) Approach control services.

c) Airport Advisory Service.

Answer – c. An Airport Advisory Service is a service like an ATCT in that they are both staffed. Unlike an ATCT the Airport Advisory Service, like the name suggests, is merely serving pilots through advisories of traffic patterns, winds and other reported traffic in the area but cannot give vectors and directives. Resource: AIM

47. The radius of a (generic designed) Class C airspace is typical

a) 5 NM.

b) 10 NM.

c) 20 NM.

Answer – b. The FAA constructs a "generic" brand of Class C airspace, which is 10 NM in total. The first ring of Class C is 5 NM and the second is another 5 NM making a total of a 10 NM radius. Class C (and Class B and D) can be constructed custom to accommodate for local features such as a cut-out (a.k.a. keyhole) for an airport within the airspace or to follow the local topography such as a river. Physical features like a river make it easier for pilots to identify the boundary of the airspace. Reference: AIM

48. All UA flight operations within Class C airspace must be

a) on a flight plan filed before the mission start time.

b) approved before the mission by the Class C Air Traffic Control Tower (ATCT) in which the flight operation will take place.

c) cleared with the Remote Pilot (RP) 24 hours before the mission time for the flight to be submitted to an FAA Operations Inspector.

Answer – b. All flights in Class C (B and D) airspace are air traffic controlled airspace and require prior permission in order to do a flight mission via the FAA waiver request available athttps://www.faasafety.gov/gslac/ALC/course_content.aspx?cID=42&sID=505&preview=true. Reference: CFR 107.41

49. The vertical limit of Class C airspace above the primary airport is normally

a) 1,200 feet AGL.

b) 3,000 feet AGL.

c) 4,000 feet AGL.

Answer – c. The FAA has a "generic" layout of Class C airspace, which is 10 NM radius wide and 4,000' AGL in height. Custom dimensions can be designed by the FAA to accommodate for local features like a nearby airport or topography. Reference: AIM

50. Under what condition may a UA aircraft operate from a non-controlled satellite airport within Class C airspace?

a) The Remote Pilot (RP) must file a flight plan before departure.

b) The UA flight crew must monitor the Class C, ATC while flight operations at the satellite airport are taking place.

c) The Remote Pilot (RP) must attain permission from Class C, ATC via the FAA online waiver request form, and notify the non-towered satellite airport management.

Answer – c. Since the question states the non-towered airport is within the Class C airspace, then the RP would be required to gain prior permission from the Class C ATCT via the FAA online waiver request: https://www.faasafety.gov/gslac/ALC/course_content.aspx?cID=42&sID=505&preview=true, and the manager/owner of the non-towered airport. Note: if the non-towered airport located in a cutout, or keyhole, of Class C then the RP would only have to notify the owner/manager of the non-towered satellite airport. However, it would be wise to gain authorization from the Class C ATCT as a safety precaution. Reference: CFR 107.41.

51. What prior action must a Remote Pilot (RP) take before starting a flight mission in Class C airspace?

a) Contact the Class C ATCT via the FAA online waiver request form to gain permission to operate in Class C airspace.

b) Contact the Class C approach control upon take-off.

c) Contact the FSS for traffic advisories.

Answer – a. Flying in Class C airspace required prior permission of the controlling agency, in this case, the Class C ATCT. The FAA request pilot does not contact the ATCT directly but use the online waiver request form available at:

https://www.faasafety.gov/gslac/ALC/course_content.aspx?cID=42&sID=505&preview=true.
Reference: CFR 107.41

52. To accomplish a UA flight mission in a Terminal Radar Service Area (TRSA) the Remote Pilot (RP) must

a) understand that operations in this area are voluntary and permission not required.

b) contact the military branch that has control over the airspace.

c) contact the TRSA, ATCT for permission for flight operation before flight via the FAA waiver request available online.

Answer – c. TRSA's (pronounced Ter-Sa) is a throwback to the old airspace classification system. Due to the unique nature of the TRSA they remained in the new system of what many call the "alphabet" airspace classification system. While they look large and have an associated ATCT, they typically have lower traffic levels. Never-the-less, the same pre-permission protocol must be followed by the RP by filling out the online FAA waiver request form at:

https://www.faasafety.gov/gslac/ALC/course_content.aspx?cID=42&sID=505&preview=true.
Reference: CFR 107.41

53. When must a current Remote Pilot (RP) certificate be in the RP's personal possession or readily accessible?

a) When acting as office manager of a commercial flight aerial photography business utilizing a UA.

b) Only when waivers have been signed by actors and extras on a movie set for a commercial drone operation.

c) Anytime when acting as the RP of a commercial drone flight operation.

Answer – c. Just as with driving an automobile in the US, the RP is required to have his/her license in their possession that can be presented to the FAA or Law Enforcement. The expression commonly used is that the RP certificate (license) is required to be "on his or her person," meaning in his or her possession. Reference: CFR 107.7

54. What document(s) must be in your personal possession or readily accessible on the flight mission site while operating as a Remote Pilot (RP)?

a) Certificates showing accomplishment of a UA checkout by an FAA flight instructor and a current biannual flight review.

b) A Person Manipulating the Controls (PMC) certificate with an endorsement of an annual flight review and a logbook showing recency of flight experience.

c) FAA issued RP certificate.

Answer – c. Just as with driving an automobile in the US, the RP is required to have his/her license in their possession that can be presented to the FAA or Law Enforcement. The expression commonly used is that the RP certificate (license) is required to be "on his or her person," meaning in his or her possession. Reference: CFR 107.7

55. Each person who holds a Remote Pilot (RP) certificate shall present it for inspection upon the request of the FAA Administrator, the National Transportation Safety Board (NTSB), or any

a) authorized representative of the National Air Traffic Controllers Association (NATCA).

b) person in a position of authority.

c) federal, state, or local law enforcement officer.

Answer – c. The FAA Administrator means anyone acting on behalf of the Administrator such as an FAA employee, like an Operations Inspector. While the NTSB is over, most manned aircraft accidents the FAA has taken a lead role in accident reporting and investigation of UA. However, due to the historic role of the NTSB makes airmen certificates subject to inspection. Since law enforcement (local, state or federal), are typically first to respond to aircraft accidents (or other issues) they are authorized to inspect an RP's certificate. CFR 107.7, 61.3

56. Under what condition, if any, may a sUAS flight crew fly through a Restricted or Prohibited area?

a) When flying on airways with an ATC clearance.

b) With permission from the controlling or using agency.

c) The regulations do not allow flight in restricted and prohibited areas.

Answer – b. By consulting the Legend of aeronautical charts, an RP can find the controlling agency over Restricted and Prohibited Airspace. Permission must be granted before flight. Gaining permission to fly in Restricted airspace will be much easier than Prohibited. Manned aircraft are, as the term implies, prohibited to fly in Prohibited airspace, so the possibility of gaining access is extremely limited. Prohibited areas are were items of national security are located such as; Washington, DC, and sensitive military areas. Restricted areas are typically military operations areas in which, with prior planning, can be accessed with appropriate planning. Reference: CFR 107.45

57. Which of the following are conditions that may be covered under an NOTAM as a TFR?

a) Presidential movements, major sporting events, disaster areas and aerial demonstrations. Sandstorms and expansive areas of thunderstorms are also covered under weather phenomenon areas.

b) Presidential movements, disaster areas, periods of unusually high barometric pressure conditions and increases in stall speeds due to high-density altitude conditions.

c) Presidential movements, major sporting events, disaster areas, aerial demonstrations and periods of unusually high barometric pressure conditions.

Answer – c. Most flight crews associate a Temporary Flight Restriction (TFR) with Presidential movements. However, they also include major sporting events, disaster areas, aerial demonstrations (such as airshows) and periods of high barometric pressure conditions. Areas where large numbers of spectators is a common sense type of safety concern. Disaster areas typically contain numerous aerial relief efforts where general flight operations (or disaster sight-seeing or reporting) may hamper relief operations. Access to disaster areas can be obtained, but is very limited. "Unusually high barometric settings" is an unusual outlier. The reason for this type of TFR is due to the maximum and minimum settings available on a standard manned aircraft altimeter. If the altimeter cannot be set because of the limitations of the instrument, then flight operations are temporarily suspended. Reference: CFR 107.47, 91.137-145 and 99.7

58. When flying sUAS in the vicinity of an airport, heliport, or seaplane base the Remote Pilot (RP) is required to

a) operate in a manner that does not interfere with operations and traffic patterns.

b) operate with the knowledge that sUAS have the right-of-way over all other aircraft.

c) make the sUAS visible to aircraft in the traffic pattern by flying at the 3 o'clock position of a manned aircraft on downwind.

Answer – a. sUAS flight crews must give-way to all other aircraft operations. CFR 107.43

59. Except in Alaska, during what time should lighted position lights be displayed on a sUAS?

a) Lighting systems are not required during times of civil twilight due to the amount of light available well after official sunset.

b) During periods of civil twilight with an anti-collision lighting system that is visible for at least 3 statute miles. The Remote Pilot (RP) may reduce the intensity of the lighting system if it is in the interest of safety to do so considering operating conditions.

c) During periods of civil twilight with an anti-collision lighting system that is visible for at least 5 statute miles. The Remote Pilot (RP) may raise the intensity of the lighting system as civil twilight expires. Additional lighting intensity may enhance midair collision avoidance safety.

Answer – b. According to TimeAndDate.com, "Civil twilight occurs when the Sun is between 0 degrees and 6 degrees below the horizon. In the morning, civil twilight begins when the Sun is 6 degrees below the horizon and ends at sunrise. In the evening, it begins at sunset and ends when the Sun reaches 6 degrees below the horizon." I like to tell my flight students that it's that time of just before or after official sunrise and after official sunset when it is still "good" light but not dark, called civil twilight. Anti-collision lights, visible up to 3 SM, are required during times of civil twilight but can be reduced in intensity if deemed safe by the RPIC. Reference: CFR 107.29, website link:

http://www.timeanddate.com/astronomy/different-types-twilight.html.

60. You arrive at the flight mission site 40 minutes before official sunrise. How long must you wait before flight operation can begin?

a) 40 minutes

b) 10 minutes

c) 1 hour 10 minutes

Answer: a. I don't like this question because it doesn't clarify if the aircraft has an anti-collision lighting system, visible to 3 SM. In this question the FAA doesn't imply there is a lighting system, this means the flight operation is limited and can't begin until official sunrise. Resource: CFR 107.29

61. You arrive at the flight mission site 40 minutes before official sunrise. How long must you wait before flight operation can begin? You have an aircraft that is outfitted with anti-collision light visible to 3 SM.

a) 40 minutes

b) 10 minutes

c) 1 hour 10 minutes

Answer – b. This question is phrased better because it specifies that anti-collision lights are installed, which means the flight operation can begin 30 minutes before official sunrise (civil twilight). Considering this the crew would have to wait 10 minutes before starting flight operations.

62. Which statement about longitude and/or latitude is true?

a) Lines of longitude are parallel to the Equator.

b) Lines of longitude cross the Equator at right angles.

c) The 0° of latitude passes through Greenwich, England.

Answer – b. Memory aid, lines of longitude run long (North -South) across the Earth. Or as in the answer, or at right angles to the Equator. Reference: PHAK

63. Figure 22, Area 3: What is the latitude and longitude of Shoshone County Airport.

a) 47° 20 N - 116° 32 W.

b) 47° 33 N – 116° 11 W.

c) 47° 45 N – 116° 40 W.

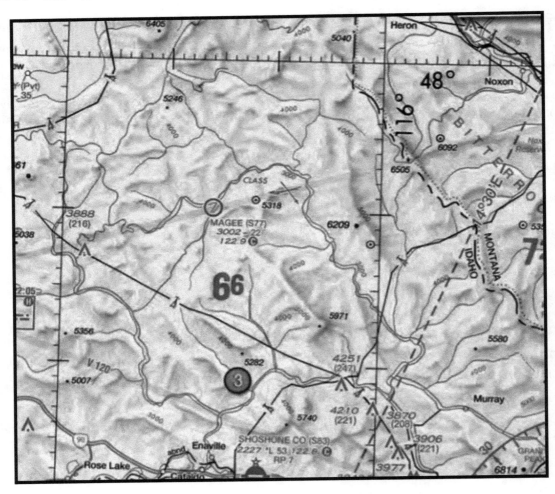

Figure 22, Area 3. Not to scale.

Answer – b. Remember that latitude/longitude (lat/long) lines are referenced in degrees separated in 60-minute intervals. After locating a reference lat/long near the location in question the next reference line will be blank (without a number), this is the 30-minute point. A common mistake is to think this blank line is the next whole degree line. Also remember, lines of latitude (East-West lines) become larger in number when going from South to North (US example). Lines of longitude go up in number when going from East to the West. Reference: PHAK

64. Figure 22: What airport is located at 47° 34 minutes 00 seconds N latitude and 100° 55 minutes 00 seconds W longitude?

a) Makeeff.

b) Poleschook.

c) Fischer.

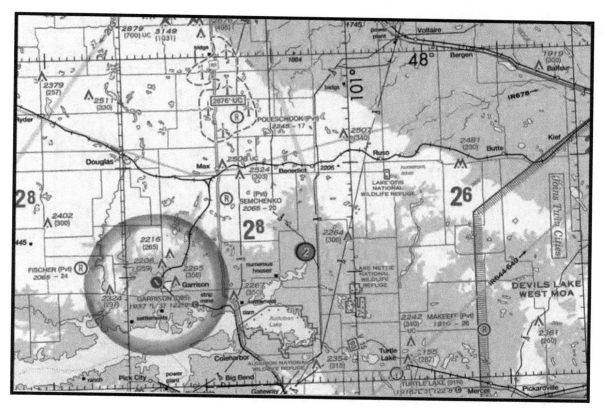

Figure 22. Not to scale.

Answer – a. Remember that latitude/longitude (lat/long) lines are referenced in degrees separated in 60-minute intervals. After locating a reference lat/long near the location in question the next reference line will be blank (without a number), this is the 30-minute point. A common mistake is to think this blank line is the next whole degree line. Also remember, lines of latitude (East-West lines) become larger in number when going from South to North (US example). Lines of longitude go up in number when going from East to the West. Reference: PHAK

65. Figure 21: What airport is located at 47° 53' 00" N latitude and 101° 15' 00" W longitude?

a) Johnson.

b) Makeeff.

c) Poleschook.

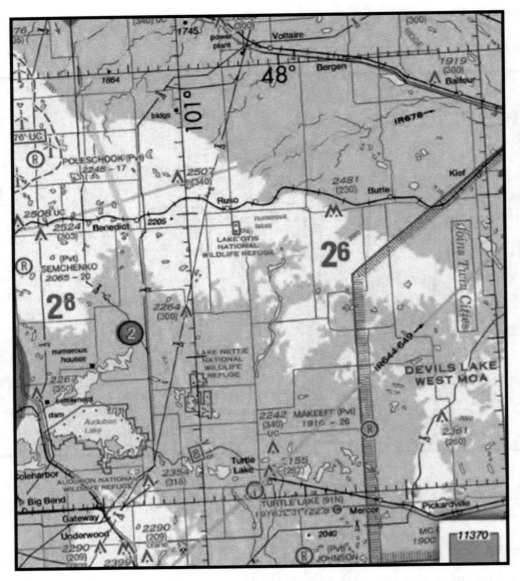

Figure 21. Not to scale.

Answer – b. Remember that latitude/longitude (lat/long) lines are referenced in degrees separated in 60-minute intervals. After locating a reference lat/long near the location in question the next reference line will be blank (without a number), this is the 30-minute point. A common mistake is to think this blank line is the next whole degree line. Also remember, lines of latitude (East-West lines) become larger in number when going from South to North (US example). Lines of longitude go up in number when going from East to the West. Reference: PHAK

66. Figure 21: What airport is located at 47° 21'N latitudes and 101° 01'W longitude?

a) Underwood.

b) Washburn.

c) Pietsch.

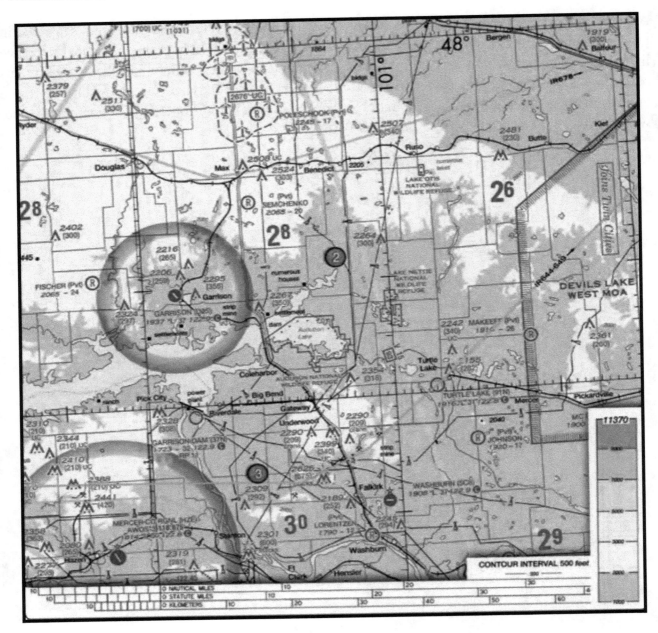

Figure 21. Not to scale.

Answer – b. Remember that latitude/longitude (lat/long) lines are referenced in degrees separated in 60-minute intervals. After locating a reference lat/long near the location in question the next reference line will be blank (without a number), this is the 30-minute point. A common mistake is to think this blank line is the next whole degree line. Also remember, lines of latitude (East-West lines) become larger in number when going from South to North (US example). Lines of longitude go up in number when going from East to the West. Reference: PHAK

67. Figure 26, Area 2. What are the latitude and longitude of the Cooperstown Airport?

a) 47° 55N - 97° 06W.

b) 47° 25N – 98° 06W.

c) 47° 55N – 98° 06W.

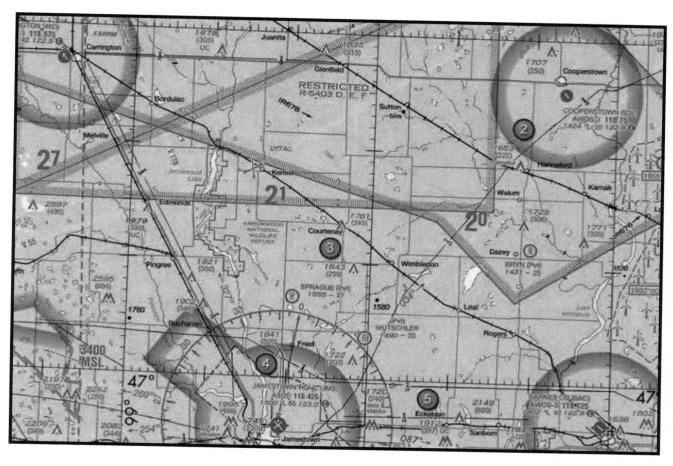

Figure 26, Area 2 (Not to scale).

Answer – b. Remember that latitude/longitude (lat/long) lines are referenced in degrees separated in 60-minute intervals. After locating a reference lat/long near the location in question the next reference line will be blank (without a number), this is the 30-minute point. A common mistake is to think this blank line is the next whole degree line. Also remember, lines of latitude (East-West lines) become larger in number when going from South to North (US example). Lines of longitude go up in number when going from East to the West. Reference: PHAK

68. Figure 20, Area 1. What is the latitude and longitude of Norfolk International Airport?

a) 36° 54N – 76° 11W.

b) 36° 20N – 76° 00W.

c) 37° 05N – 75° 11W.

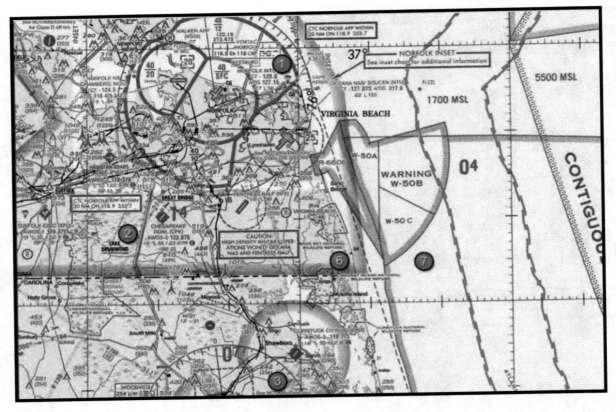

Figure 21 Area 1. Not to scale.

Answer – a. Remember that latitude/longitude (lat/long) lines are referenced in degrees separated in 60 minute intervals. After locating a reference lat/long near the location in question the next reference line will be blank (without a number), this is the 30-minute point. A common mistake is to think this blank line is the next whole degree line. Also remember, lines of latitude (East-West lines) become larger in number when going from South to North (US example). Lines of longitude go up in number when going from East to the West. Reference: PHAK

69. Figure 78 and Figure 79. What frequency at the Sioux Gateway/Col Day Airport be monitored when a sUAS flight crew has a mission in the airport area at 1100Z?

a) 122.95 MHz.

b) 119.45 MHz.

c) 118.7 MHz.

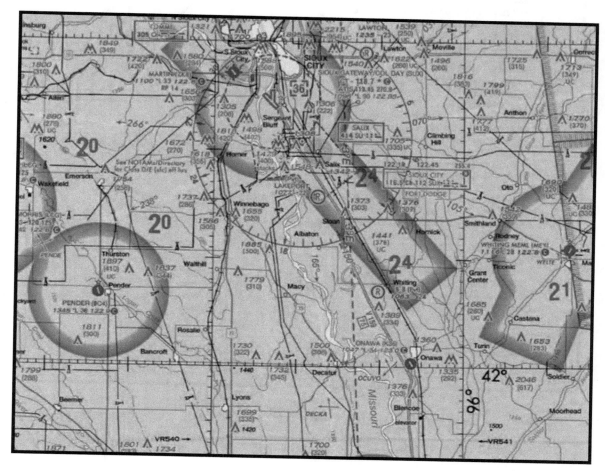

Figure 78 (Not to scale).

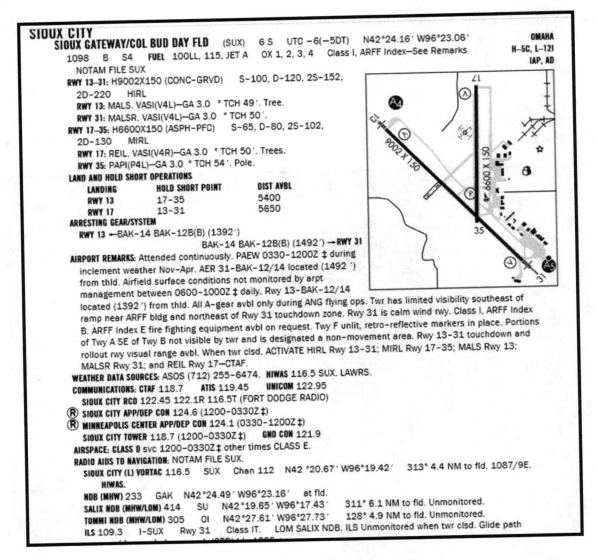

Figure 79

Answer – c. You don't necessarily need to find the Sioux City Gateway/Col Day Airport (KSUX) with the lat/long, but it is provided along with the Airport Supplement information. After ATCT hours UA flight crews can monitor CTAF, the same as the Tower frequency when the tower is closed. Details about the airport can be found in detail on the referenced Chart Supplement page (Figure 79). Under "Airspace" it states that the KSUX airport is "Class D (SVC 1200-0330Z) other times Class E." Note: The Airport Supplement can be "decoded" by using the information included (but not directly referenced) in the Knowledge Test Supplement available at http://www.faa.gov/training_testing/testing/test_questions/media/sport_rec_private_akts.pdf (also includes other Legends and Figures). Reference: PHAK, Airport Supplement, Aeronautical chart

70. Which is true concerning the blue colors used to depict airports (a representation of the runways) on Sectional Aeronautical Charts?

a) Airports with control towers underlying Class A, B, and C airspace are shown in blue; Class D and E airspace are magenta.

b) Airports with control towers underlying Class C, D, and E airspace are shown in blue.

c) They are airports with Air Traffic Control Towers.

Answer – c. Blue is typically used for airports that have ATCT's. Magenta (a red-ish color) depicts airports without an ATCT. Reference: AIM

71. Class E airspace extends upward from either 700 feet or 1,200 feet AGL, but does not include,

a) 12,000 feet MSL.

b) 14,500 feet MSL.

c) 18,000 feet MSL.

Answer – c. Class G airspace extends from the ground to the base of Class E airspace at 700' or 1200' AGL. Where is Class F airspace? There is "no F airspace," and that's the closest memory aid I can give you in this family friendly study guide. The airspace graphic below can give you a good basic idea of airspace types. Reference: AIM

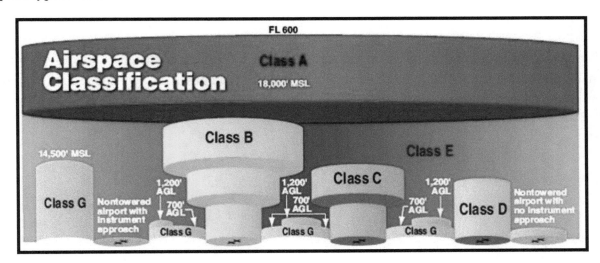

72. Figure 21, Area 2. What type of airspace is the Fentress NALF (NFE) Airport located?

a) Class E.

b) Class G.

c) Class C.

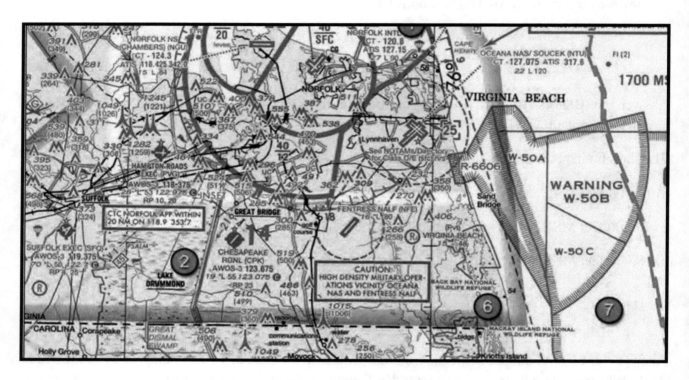

Figure 20 Area 2

Answer – a. After locating the airport on the aeronautical chart reference the chart legend (provided in the materials) to determine the type of airspace. KNFE has a magenta dashed line around the airport meaning it is Class E starting at the surface (a.k.a. surface based Class E airspace). Reference: AIM

73. Figure 20: What hazards to aircraft may exist in Restricted Areas R-5302A, located in the Albemarle Sound area?

a) Military training which can include invisible hazards such as aerial gunnery or guided missiles.

b) The high volume of civilian pilot training.

c) Numerous amphibious civilian aircraft operations.

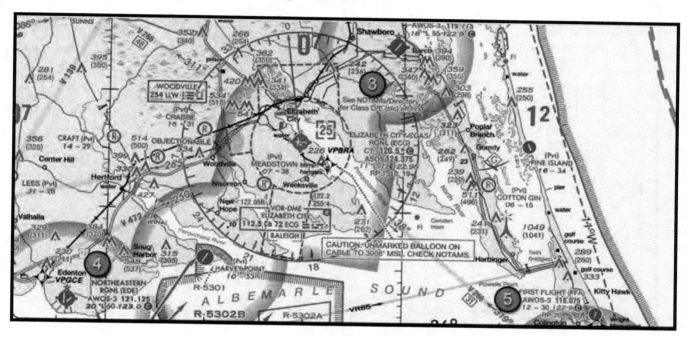

Figure 20 (Not to scale).

Answer – a. Restricted areas are set aside areas in which the military holds live-fire exercises (air to ground and ground to air), the danger is a lot higher than other military training areas. They differ from Military Operations Areas (MOAs) in that MOAs are typically used for general military training exercises. Restricted Areas have more stringent access requirements such as the need for prior permission from the controlling agency. The altitudes, times of effective use and the controlling agency can be found on the aeronautical chart legend. Reference: AIM, aeronautical chart legend.

74. Figure 21, Area 2: What is the elevation of the Fentess NALF Airport?

a) 80 feet.

b) 800 feet.

c) 16 feet.

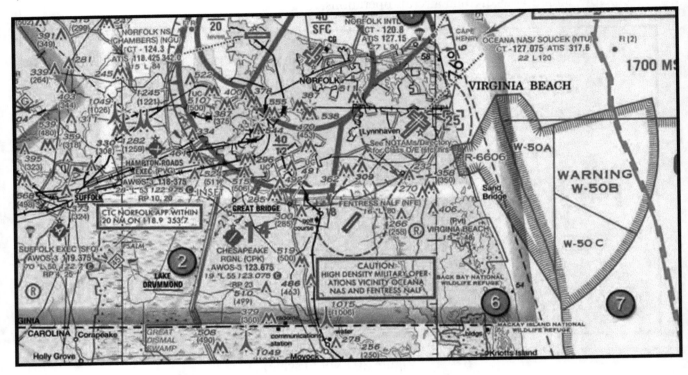

Figure 20, Area 2 (Not to scale).

Answer – c. As a general reference, you might guess that Fentess NALF is close to Sea Level, and you would be correct. To get the specific elevation find the airport information block on the chart where it states the name of the airport. There are two lines of information, and the second line gives; elevation, lights and length of runway respectively. The first number of the third line is the elevation, which is 16' MSL. By the way, there isn't a frequency listed because this is a Naval Aviation Landing Field (NALF). Reference: Chart Legend

75. When planning for a sUAS flight mission, you notice part of the mission area is in a National Park Service (NPS) area. What does this mean for flight operations in this area?

a) No further authorization is necessary for operations over NPS areas. Since the mission originates off of NPS land, the flight can proceed as planned.

b) The NPS has prohibited the use of sUAS in NPS park lands. They will permit, under specific NPS approval, the use of aircraft for emergency situations such as disasters and Search and Rescue (SAR).

c) Authorization is necessary from the mayor of the closest city associated with the specific NPS park. A fee of $50 must be paid directly to the city tax and fee office 48 hours before the flight mission.

Answer – b. The federal NPS has prohibited drone use within the park system, citing the disruption of visitors and wildlife. There are very limited instances in which permission can be gained, such as natural disasters and SAR operations. State parks are a different issue depending on the State. Check your State Park systems website for specifics. Reference: 36 CFR 2.17(a)(3), website link:

https://www.nps.gov/yose/learn/news/use-of-unmanned-aircraft-systems-drones-prohibited-in-yosemite-national-park.htm

76. What action should a Remote Pilot (RP) take when operating in a Military Operations Area (MOA)?

a) Obtain a clearance from the controlling agency before entering the MOA.

b) Operate only on the airways that transverse the MOA.

c) Exercise extreme caution when military activity is being conducted.

Answer – c. During a preflight check of the weather the RP can ask the weather briefer (1800-WXBRIEF or 1800WXBRIEF.com) if a specific MOA is "hot" (active) or "cold" (inactive). While the answer states UA operations can take place by exercising extreme caution, we suggest contacting the controlling agency (found on the aeronautical chart legend). The chart legend will also show altitudes and the controlling agency information. Reference: AIM

77. Flight through a Restricted Area should not be accomplished unless the Remote Pilot (RP)

a) has filed a mission plan with the Automated Flight Service Station (AFSS).

b) received prior authorization from the controlling agency.

c) received prior permission from the commanding officer of the nearest military base.

Answer – b. The restrictiveness of airspace, from most restrictive to the least, is Prohibited (no normal access), Restricted (limited), Military Operations Area (open with extreme caution if "hot" or active) and Alert Area (open with vigilance) followed by Military Routes (Visual Routes and Instrument Routes) that are open with vigilance. The aeronautical chart legend will give more information at the altitudes, and controlling agency contact for the specific Restricted Area used to gain permission. Reference: Aeronautical chart legend, AIM

78. Figure 21, Area 1 and Area 2. The terrain elevation of the light tan area (scroll down further to see the second graphic) between Minot (Area 1) and Audubon Lake (Area 2) varies from

a) 2,800 feet to 3,200' MSL.

b) 2,000 feet to 2,676' MSL.

c) Sea Level to 2,000' MSL.

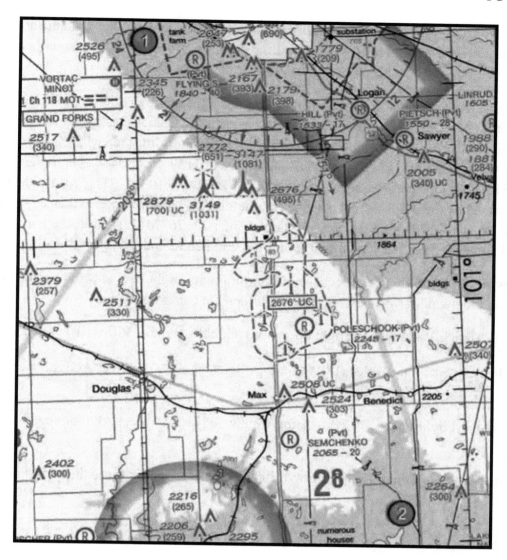

Figure 21, Areas 1 and Area 2. Not to scale.

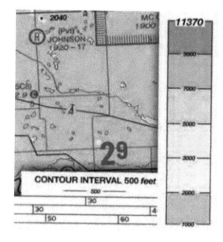

Figure 21 Elevation Contour Color Chart

Answer – b. This question can be a little tricky unless you look closely at the "light tan area" elevations. First, look at the elevation color chart and find the light tan block. Light tan covers elevations from 2,000' MSL to 3,000' MSL, so these altitudes must be the answer.

Look for high elevation areas by focusing only on the light tan areas between Minot (Area 1) and Audubon Lake (Area 2). The wind farm (reference chart legend) has the number 2676' Under Construction (UC). Look for other high altitude areas. Deceivingly, the area where a large tower located North of the wind farm looks to be the highest in the area at 3149' MSL. However, remember the tower itself, at 1031' AGL, must be subtracted from 3149' MSL. 3149 – 1031 = 2118' MSL. The set of double towers Northwest of the 3148' MSL tower set on the highest terrain in the area at 2179' MSL (2879' MSL – 700' AGL = 2179' MSL). The wind farm wins at 2676' MSL (note there is no AGL turbine height listed since the farm has structures, are minimal in height under construction).

The best answer is "b" in this case because of the minimum "light tan" elevation reference begins at 2000' MSL and the wind farm is at 2676' MSL. Reference: Chart Legend

79. Figure 21, Area 3: What type of flight operations should a Remote Pilot (RP) expect along IR 644?

a) Military Instrument Flight Rules (IFR) training flights above 1,500 feet AGL at speeds more than 250 knots.

b) Military Visual Flight Rules (VFR) training flights above 1,500 feet AGL at speeds less than 250 knots.

c) Military Instrument training flights below 1,500 feet AGL at speeds more than 150 knots.

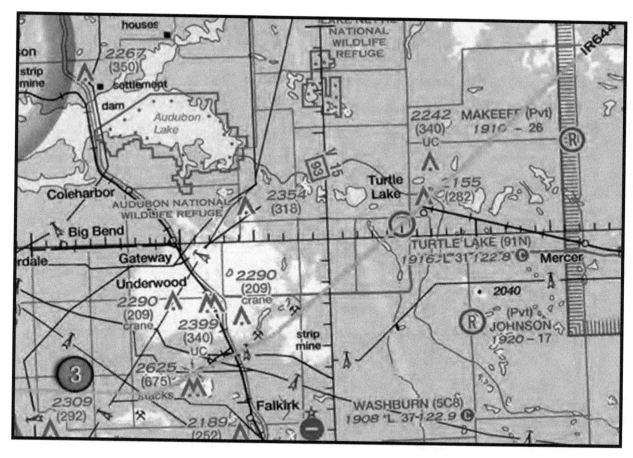

Figure 21, Area 3 (Not to scale).

Answer – a. The gray lines (think gray military aircraft) beginning below Area 3 and continuing West to Northeast are Instrument (military training) Routes (IR) and begin at 1500' AGL. Notice IR 644 in the very top right-hand corner of the chart. Needless to say, these are fast moving aircraft. A memory aid is to remember that Instrument training requires in-cockpit focus relying solely on the instruments. As a result, Instrument flying has to be accomplished at a higher altitude than their Visual Route (VR) counterparts. Note: VR can be high or low altitude. Reference: AIM

80. To obtain more information concerning a parachute jumping symbol seen on an aeronautical chart consult the

a) FAA Air Routing Traffic Control Center (ARTCC).

b) Aeronautical Chart Supplements.

c) Graphic Notices and Extraneous Data.

Answer – b. An aeronautical chart will show the parachute symbol as a general notice of such operations at an airport. To find more specific information an RP should consult the Chart Supplement for the specific airport associated with the parachute symbol. Reference: Chart Supplements (previously known as the AFD).

81. Figure 22, Area 3: The Class E airspace designated as a Federal Airway over Magee Airport extends from

a) 1,200 feet AGL to 17,999 feet MSL.

b) 700 feet MSL to 12,500 feet MSL.

c) 7,500 feet MSL to 17,999 feet MSL.

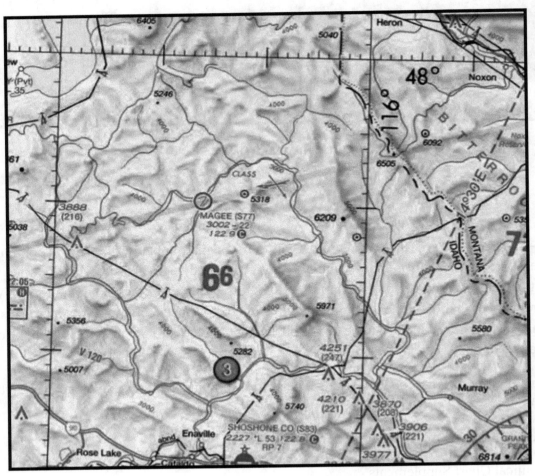

Figure 22, Area 2. Not to scale.

Answer – a. The light blue line that crosses Magee is a Victor Airway (a highway in the sky for manned aircraft). Don't forget you have access to the chart Legend to help determine it is an airway. Airways begin at 1,200' MSL and extend to the base (17,999' MSL) of Class A airspace at 18,000,' or more correctly said, Flight Level 180 (one-eight-zero). Now you can talk like an airline captain.

Looking at similar depictions in other areas of the chart may help to jog your memory. Notice there is another Victor Airway, V 120, below the Area 3 symbols. This shows the airway number, V 120, unlike the one over Magee. Reference: PHAK, AIM, Chart Legend.

82. Figure 21, Area 1, look Northwest of Norfolk Intl Airport (KORF), inside of Class C. For information about the parachute jumping Northwest of KORF a pilot can consult the

a) Controlling Agency listed on the border of the aeronautical chart.

b) The chart Supplements, formerly known as the Airport Facility Directory.

c) The specific States Airport Guide.

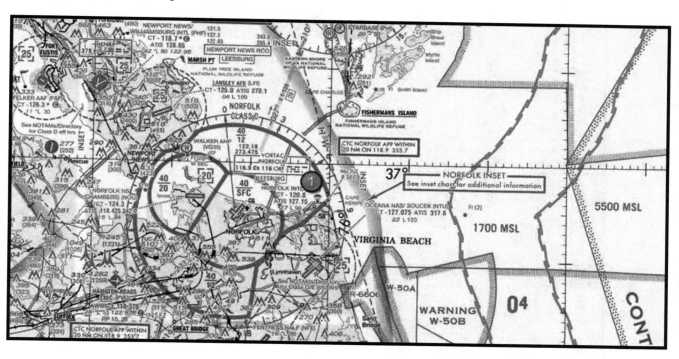

Figure 21, Area 1. Not to scale.

Answer – b. Look, just below the number 1 for the parachute symbol. The parachute symbol (cross-check symbols with the Chart Legend if needed) on the aeronautical chart just indicates there are jump operations at the airport. For more information, consult the Chart Supplements. Reference: Chart Legend, Chart Supplement, PHAK, AIM

83. Figure 25, Area 2: What is the base and ceiling of the Dallas Fort Worth Intl Airport (KDFW) Class B airspace over the Air Park-Dallas Airport (F69)?

a) 2,000 MSL to 11,000' MSL.

b) 3,000' MSL to 11,000' MSL.

c) 3,000' AGL to 11,000' AGL.

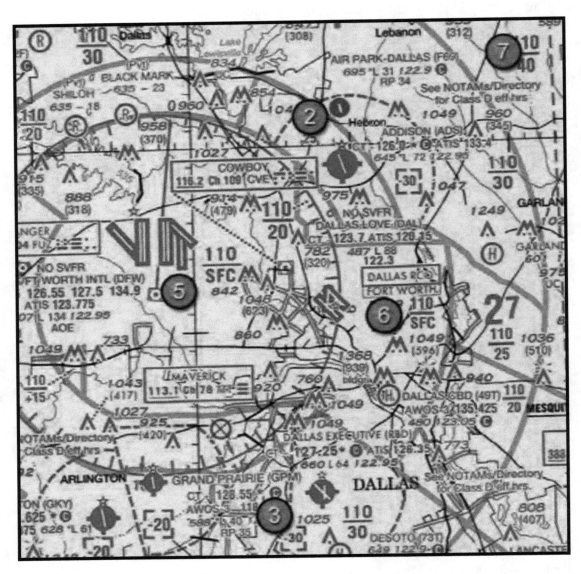

Figure 25, Area 2. Not to scale.

Answer – b. The F69 Airport is the small non-air traffic controlled (magenta) airport adjacent to the number 2. Remember: Class B airspace can be visualized like an upside-down wedding cake. The light blue rings show the KDFW Class B. Inside of each ring is an altitude figure. The closer a person is to KDFW the lower to the ground Class B will extend, for example KDFW is from the Surface (SFC) to 11,000' MSL (see figure next to Area 5). Follow the ring over F69 to the upper left-hand section of the graphic and you will see 110/30. In aviation, there are many instances in which two zeros are added, such as in the example. Therefore, the upper limit of Class B is 11,000' MSL, and the lower limit is 3000' MSL. Remember: Altitudes are given primarily in MSL because manned aircraft pilots use altimeters that use MSL as a reference. Some people called MSL as good "working number" for manned aircraft. Reference: Aero Chart, CFR, AIM, PHAK

84. Figure 25, Area Southwest of Dallas-Fort Worth Intl (KDFW). What type of airspace is Fort Worth Spinks Airport (KFWS) and how high does it extend? Additionally, what are the altitudes that the KDFW Class B airspace above KFWS?

a) KFWS is considered Class B airspace to 11,000 feet MSL because it is located under the shelf, on the KDFW Class B.

b) KFWS is Class C airspace from the Surface (SFC) to 5,000' feet MSL. KDFW Class B extends from 4,000' MSL to 10,000' MSL.

c) KFWS is Class D airspace from the Surface (SFC) to 3,000' feet MSL. KDFW Class B extends from 5,000' MSL to 11,000' MSL.

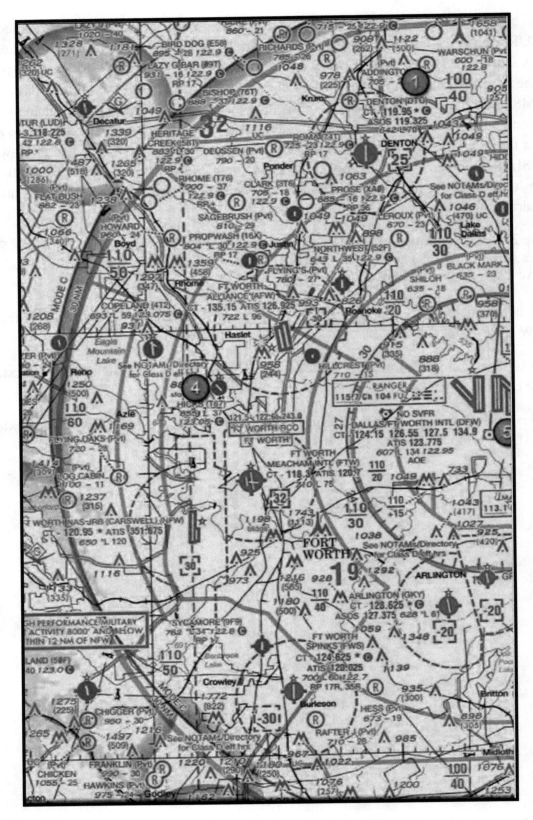

Figure 25, Southwest of Dallas-Fort Worth Intl, KDFW. Not to scale.

Answer – c. The KFWS Airport is Southwest of KDFW in Class D Airspace denoted by the blue dashed ring. To the Southwest of KFWS, there is the number 30 inside of a blue dashed square box, meaning Class D extends to 3,000' MSL. KFWS is below, but not in, KDFW Class B Airspace. Carefully following the solid light blue lines notice the 110/50 figure Northwest of KFWS over a lake. This figure means KDFW Class B extends from 5,000' MSL to 11,000' MSL in this section, which includes KFWS. Reference: Aero Chart

85. Figure 26, Area 2: What are the dimensions of the Class D Airspace at the Addison Airport?

a) Surface (SFC) to 3,000' MSL.

b) Surface (SFC) to 3,000 feet AGL.

c) 2,000' MSL to 11,000' MSL.

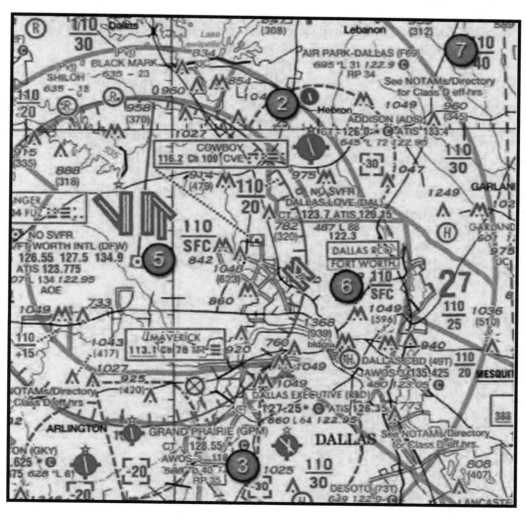

Figure 25, Area 2 (Not to scale).

Answer – b. Notice inside of the blue dashed line of Addision Class D the number 30 in a blue dashed box. This means Class D extends from the Surface (SFC) to 3,000' MSL. Reference: Aero Chart

86. To get an expanded view of Class B, a pilot should

a) view of an electronic device capable of zoom control.

b) use the Terminal Area Chart for the specific airports Class B.

c) request an access code from the Automated Flight Service Station.

Answer b. A Terminal Area Chart shows greater detail than a standard chart or by zooming in an electronic chart. Reference: PHAK

87. Figure 26, Area 7: What Class of airspace is the Collin Co. McKinney Airport (KTKI)? How high does it extend? Does it have an air traffic control tower?

a) KTKI is a Class C Airspace airport that extends from the Surface (SFC) to 2,900 feet MSL. It has a control tower whose frequency is 118.825.

b) KTKI is a Class D Airspace airport that extends from the Surface (SFC) to 2,900 feet MSL. It has a control tower whose frequency is 118.825.

c) KTKI is a Class D Airspace airport that extends from the Surface (SFC) to 2,900 feet MSL. It has a control tower whose frequency is 118.827.

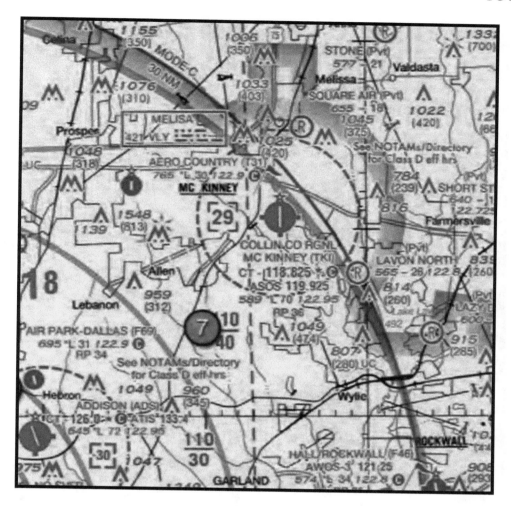

Figure 25, Area 7. Not to scale.

Answer – b. Feel free to use the airport Legend to answer this question. The blue dashed line denotes Class D Airspace. I good memory aid and the tip is to recall that blue airports have air traffic control towers. The airspace extends from the Surface (SFC) to 2,900' MSL, denoted by the number 29 in the blue box made of dashed lines just to the West of the airport. Reference: Aero Chart

88. Figure 25, Northwest of Area 4. At what altitude does the Dallas – Fort Worth (KDFW) Class B Airspace begins and end over the Copeland Airport (4T2)?

a) At the surface to Flight Level 600.

b) KDFW Class B Airspace begins at 4,000' MSL and extends to 11,000 feet MSL.

c) KDFW Class B Airspace begins at 5,000' MSL and extends to 11,000 feet MSL.

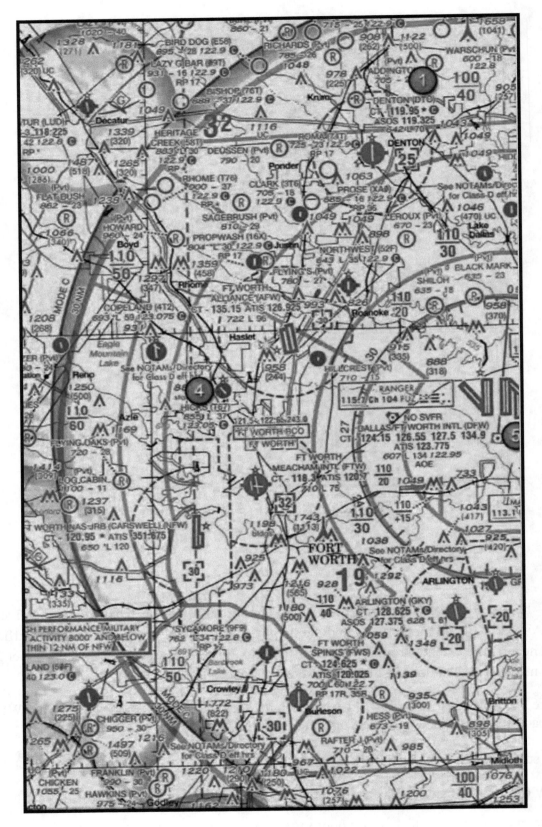

Figure 25, Northwest of Area 4. Not to scale.

Answer – c. The KDFW is one of the busiest (and complex) airports in the US, be careful to follow the blue Class B lines surrounding 4T2. Look to the Northwest of the 4T2 Airport to the numbers 50/110, meaning Class B begins at 5,000' MSL and extends to 11,000' MSL. Reference: Aero Chart, Legend

89. What type of airspace is the Devils Lake MOA, Figure 26, Area 1?

a) Myopic Operations Area (MOA).

b) Military Observation Aircraft (MOA) area.

c) Military Operation Area (MOA).

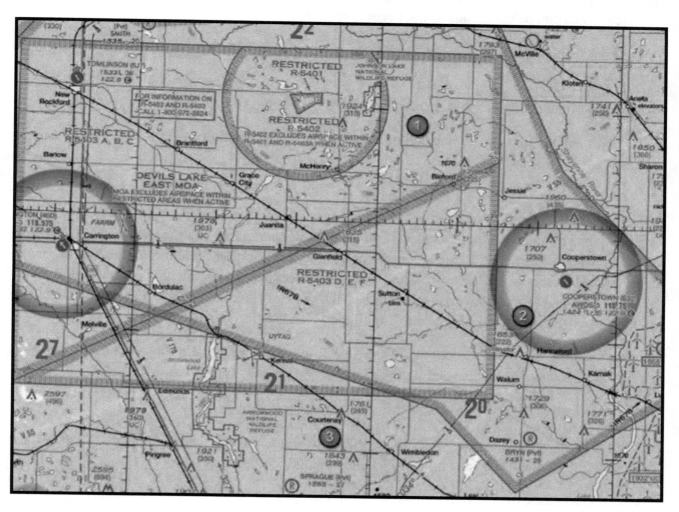

Figure 26, Area 1 (Not to scale).

Answer – c. The MOA is denoted by the magenta hatched line (reference the Chart Legend if needed) that covers a large land mass around Area 1. The complex part of this area is the "stacking" of Restricted Airspace with MOA. As you remember, MOA require extreme caution to operate, but doesn't require permission as with Restricted Areas.* Restricted Areas contain more intense and hazardous military flight operations. The perimeter of the aeronautical chart will give details (times and altitudes covered) about the Restricted and MOA areas along with the controlling agency (contact). Reference: Aero Chart, AIM.

*We suggest checking to see if an MOA is active, or "hot" via the Automated Flight Service Station (AFSS) and generally avoiding flying in this airspace during this time.

90. What type of military flight activity might be present in the Devil's Lake East MOA, Figure 26, Area 2, that can be dangerous to UA flight operations?

a) Unusual, often invisible, hazards to aircraft such as artillery, firing, aerial gunnery, or guided missiles.

b) Military training activities that necessitate acrobatic or abrupt flight maneuvers.

c) The high volume of pilot training and/or unusual type of aerial activity.

Answer – b. MOAs contain many types of military flight operations with the exception of live firing (as referenced in answer a). Pilots are encouraged to exercise extreme caution in these areas, but no permission as necessary. However, when pre-flight checking weather with the Automated Flight Service Station (AFSS) a pilot can inquire if the MOA is "hot" (active) or "cold" (inactive). If the area is "hot" it may be wise to consider waiting until the MOA is cold.

Answer c is describing more closely an Alert Area from where an initial military flight instruction is taking place, in addition to "unusual flight activities." Reference: Aero Chart, AIM

91. What Class(es) of Airspace are above Sprague Airport, Figure 26, Area 3?

a) Class G airspace extends from the Surface (SFC) up to but not including 1,200 feet AGL then Class E Airspace 1,200 feet AGL up to but not including 18,000 feet MSL.

b) Class G airspace extends from the Surface (SFC) to up to but not including 18,000 feet MSL.

c) Class G airspace extends from the surface up to but not including 700 feet MSL, then Class E airspace 700 feet to 14,500 feet MSL.

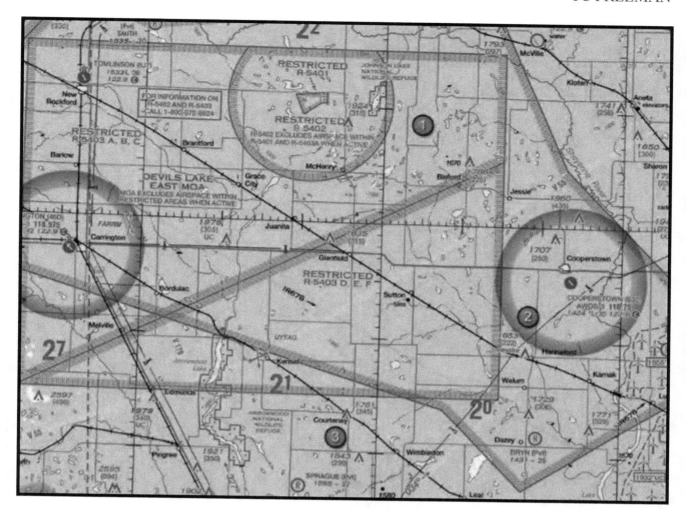

Figure 26, Area 3. Not to scale.

Answer – a. Since there is not a magenta circle Sprague Airport Class G extends from the Surface (SFC) up to, but not including 1,200' Above Ground Level (AGL). Class E extends from 1,200' AGL up to, but not including, 18,000' MSL. If you recall, Class A Airspace begins at 18,000' MSL, or more correctly, Flight Level 180 (one-eight-zero). The answers can be a bit tricky with the wording "up to but not including." Many pilots might incorrectly say, "Class G is up to 1,200 feet Above Ground Level," but who wants to say 1, 199 feet AGL" or "1,200 feet, but not including 1,200 feet AGL?" Reference: Chart Legend, AIM, CFR 71.71

92. Find the Bryan Airport, Figure 26, East of Area 3, and describe the Class(es) of Airspace over the airport.

a) Class G airspace – surface up to but not including 18,000 feet MSL.

b) Class G airspace extends from the surface up to but not including 1,200 feet AGL. Class E airspace continues from 1,200 feet AGL up to but not including 18,000 feet MSL.

c) Class G airspace extends from the surface up to but not including 700 feet MSL. Class E airspace continues from 700 feet to 14,500 feet MSL.

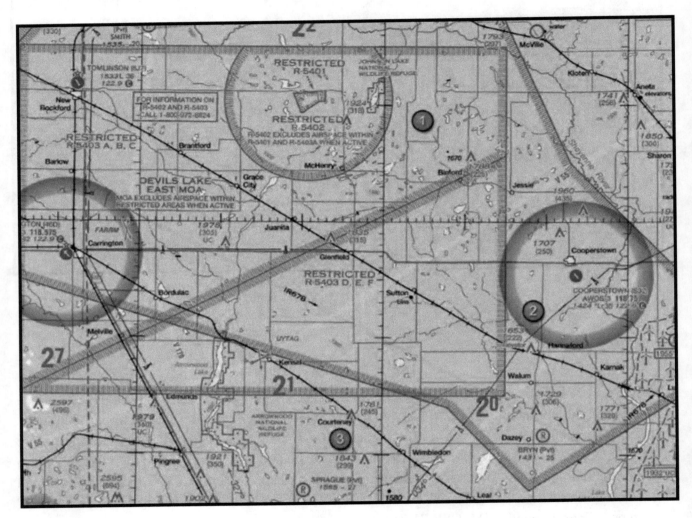

Figure 26, East of Area 3. Not to scale.

Answer – b. Since there is not a magenta circle Bryan Airport Class G extends from the Surface (SFC) up to, but not including 1,200' Above Ground Level (AGL). Class E extends from 1,200' AGL up to, but not including, 18,000' MSL. If you recall, Class A Airspace begins at 18,000' MSL, or more correctly, Flight Level 180 (one-eight-zero).

While the MOA that is over Bryan is airspace, it is considered Special Use Airspace and not considered a Class of Airspace.

The answers can be a bit tricky with the wording "up to but not including." Many pilots might incorrectly say, "Class G is up to 1,200 feet Above Ground Level," but who wants to say 1, 199 feet AGL" or "1,200 feet, but not including 1,200 feet AGL?" Reference: Chart Legend, AIM, CFR 71.71

93. What Class of Airspace is over the Barnes County Airport (KBAC), Figure 26 East of Area 5?

a) Class D airspace from the surface to the floor of the overlying Class E airspace.

b) Class G Airspace from the Surface (SFC) to 700 feet AGL.

c) Class E airspace from the surface to 1,200 feet MSL.

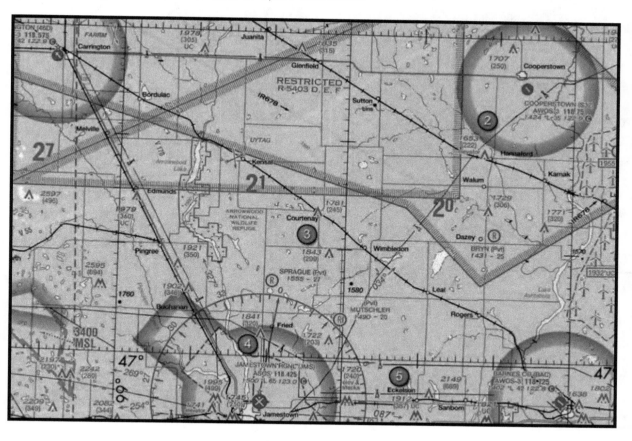

Figure 26. Not to scale.

Answer – b. Notice that the KBAC Airport has a magenta line around the airport. This means that there is an instrument approach into this airport, and the airspace designers want to protect more of the airspace than with a Visual Flight Rules (VFR) airport (clear weather only airport). Inside of the magenta circle weather requirements (minimum visibility and clearance from clouds) for manned aircraft are stricter to protect aircraft pilots from midair collisions. Outside of the magenta circle Class G extends up to, but not including 1,200' AGL. Reference: Aero Chart

94. How would you describe the Bordner Airport (3D8), Figure 59, Northeast of Area 5?

a) A private airport with a grass airstrip that has a parachute operation in the area.

b) A public airport with a runway that is not a hard surface that has a parachute operation in the area.

c) An abandoned airport that has landmark value.

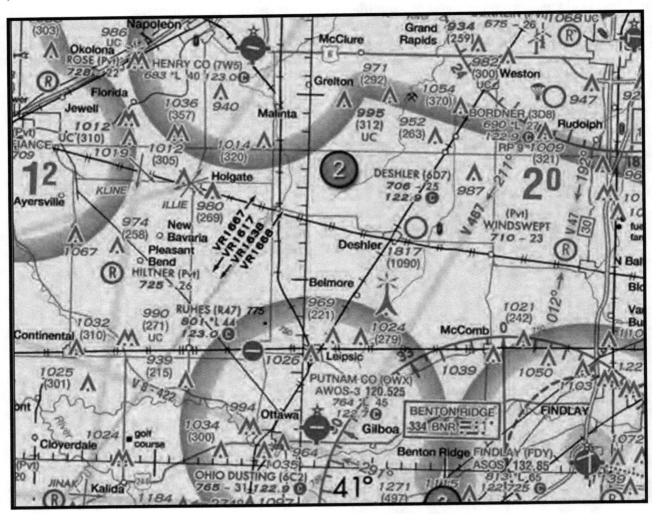

Figure 59, Northeast of Area 2. Not to scale.

Answer – b. Use the chart legend to find out the type of airport. This particular airport is open to the public and does not have a hard surface (paved) runway. Airports with an X inside of the circle are closed, but good to use as a landmark. Airports with an R inside of the circle are, Private Airports but can be used by manned aircraft for emergency purposes. Reference: Aero Chart, Legend, AIM

95. What Class of Airspace is the Findlay Airport (KFDY), Figure 59, Northeast of Area 3 and immediate surrounding area?

a) Class C Airspace.

b) Class D Airspace.

c) Class E Airspace.

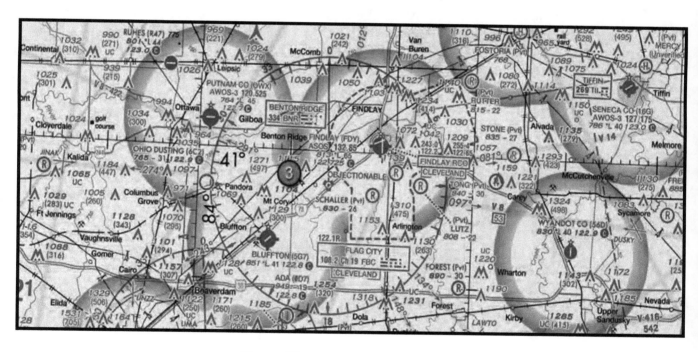

Figure 59, Area 3. Not to scale.

Answer – c. A magenta dashed line denotes Class E Airspace as indicated on the aeronautical chart legend. Notice the double keyhole shape of the Class E that align with the runways indicating instrument approaches. To protect manned aircraft from midair collisions, with stricter weather requirements (visibility and clearance from clouds), Class E begins at the Surface (SFC) and extends up to, but not including, 18,000' MSL. Reference: Aero Chart, AIM

96. Where does controlled airspace begin over the Corpus Christi Very *High-Frequency* Omni Range (VOR) near Figure 69, Area 5?

a) Surface (SFC).

b) 700 feet AGL.

c) 1,200 feet MSL.

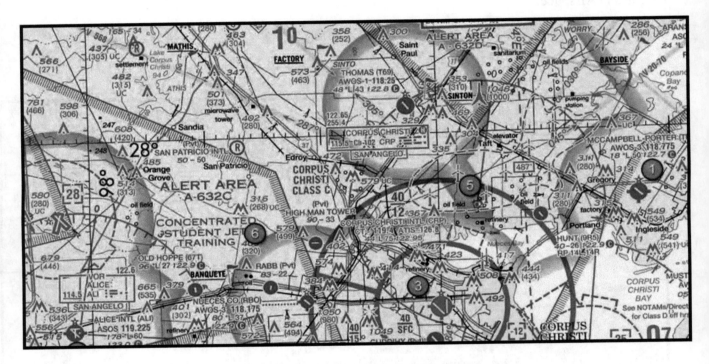

Figure 69, Area 5 (Not to scale).

Answer – b. The Very high-frequency Omni Range (VOR) symbol is in the South/Southeast of Area 5 and extends out with a large compass rose. The chart can be misleading because of the wind farm (and blue dashed boundary line), which is of no relevance to the answer. Additionally, the Corpus Christi Class C Airspace is above the VOR at 1,200' MSL. See the figure 40/12 West of the Area 5 number inside of the Class C-ring which means the airspace extends from 1,200' MSL to 4,000' MSL. The doubly tricky part is that if you carefully follow the magenta shaded line around the Thomas Airport (T69) you will see it does in fact extend down to the VOR. The reason for this extension is that the T69 uses the VOR for manned aircraft instrument approaches. This airspace has stricter weather requirements (visibility and clearance from clouds) to protect manned aircraft from midair collisions. As a result, Class E Airspace begins at 700' AGL with Class G going from the Surface (SFC) up to, but not including 700' AGL. Keep in mind Class G is uncontrolled, Class E is considered controlled. For a visual reference look at the FAA graphic below:

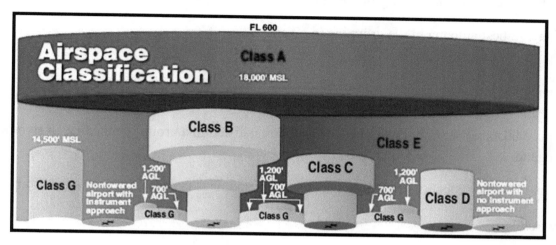

FAA Airspace Classification

Note: VORs are physical buildings (structures) that are placed around the US for manned aircraft navigation (pre-GPS). A VHF frequency is used in combination with aircraft instrumentation to guide pilots to (or from) the physical location of the building (structure) that contains the VOR. While not regularly used since the advent of GPS it remains as a good backup and for aircraft that utilize VOR aircraft components. Reference: Aero Chart, AIM

97. Antennas and obstructions can be listed on the aeronautical chart while their height is not known. How is the unknown height of the antenna or obstacle listed?

a) All antennas and obstructions on the aeronautical chart are required to have height information.

b) As an estimate, indicted by including an asterisk beside the estimated number.

c) By the letters UL.

Answer – c. An antenna or obstruction of unknown height can be listed on the aeronautical chart with the letters UL. Memory aid: Think UL as standing for Un-Listed. Reference: Aero Chart Legend

98. When a grouping of antennas is shown on the aeronautical chart why is there only one height figure?

a) This is an untrue statement, if an antenna is shown on the aeronautical chart is required to list height information.

b) Only the lowest antenna in the grouping is given.

c) Only the highest antenna in the grouping is given.

Answer: c. Makes sense, only the highest obstruction is listed in a group of antennas. Reference: Aero Chart Legend

99. The major safety concern for flight in an area with antennas is not the structure itself but the

a) guy wires.

b) lighting system.

c) legal ramifications via violation of Part 107 regulation.

Answer – a. Guy wires support the antenna structure and are difficult to see, posing a collision hazard. Manned aircraft more commonly impact a guy wire rather than the structure itself. UA crews should be aware that wires can extend out 2000' from the antenna.

100. Bobs and Sunrise Airport, Figure 71, West of Area 3 are

a) privately owned airports with restricted use.

b) An airport restricted to use by sport pilots only.

c) a restricted military airfield within restricted airspace.

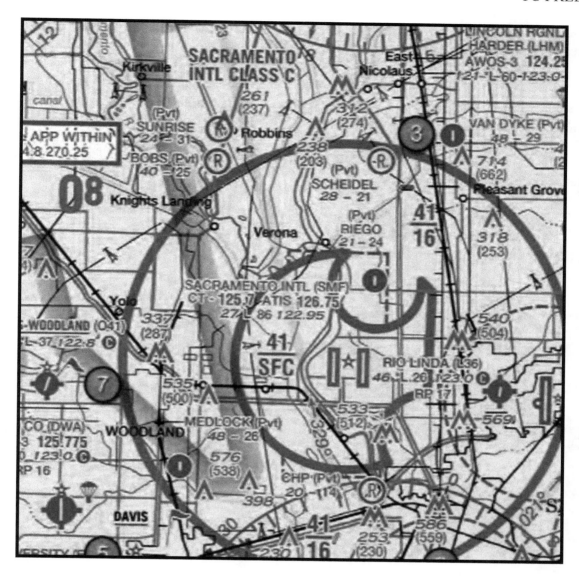

Figure 71, West of Area 3. Not to scale.

Answer – a. In the aeronautical chart Legend, you can see that a circle with an R inside of it is a Restricted Airport but not to be confused with Restricted (military) Areas. Each of the airports is privately (marked "Pvt" on the chart) owned and require prior permission from the owner to use. The only exception is for emergency situations encountered by manned aircraft pilots, hence the primary reason the airport is listed on the chart. Reference: Aero Chart, AIM

101. A customer proposes a flight mission to take pictures of rare bird species in the Wilderness Areas East and Northeast of Area 4, Figure 75. This would require

a) determination if the park is a Bureau of Land Management (BLM) runs land to determine required permissions. Most Federal and State areas do not allow UA operations, but a pilot can petition the local governmental agency, city or county to seek authorization.

b) no authorization, but that the UA flight crew is sensitive to the wildlife in the area and also the experience of the visitors of the park.

c) determination if the park is a Federal or State run lands to determine required permissions. Most Federal parks do not allow UA operations, but State areas may authorize with prior permission depending on the State and individual park.

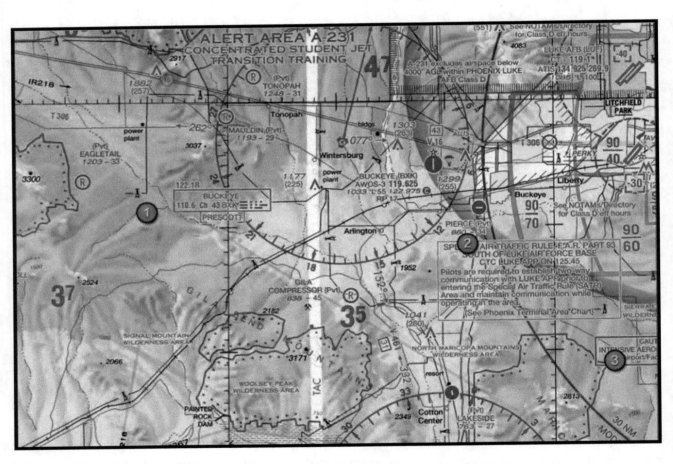

Figure 75, Wilderness Areas West of Area 3. Not to scale.

Answer – c. Wildlife areas are shown on the aeronautical chart to warn manned aircraft pilots that higher than required crossing altitudes are required for transitioning the area. Unfortunately, UA pilots are restricted from most of these areas, be it Federal or State (Federal more than State), but more research should be done to verify what permissions are necessary. Reference: National Park Service.

102. Eagles Nest Airport, Figure 71, East of Area 6 is

a) an airport restricted to use by lightweight ultralight aircraft due to sensitive paved runway conditions.

b) a restricted military airfield closed to the public that is 4,000' long, hard surface runway available for use by manned aircraft in emergency situations only.

c) a privately owned - public use 4,000' long, hard surface runway.

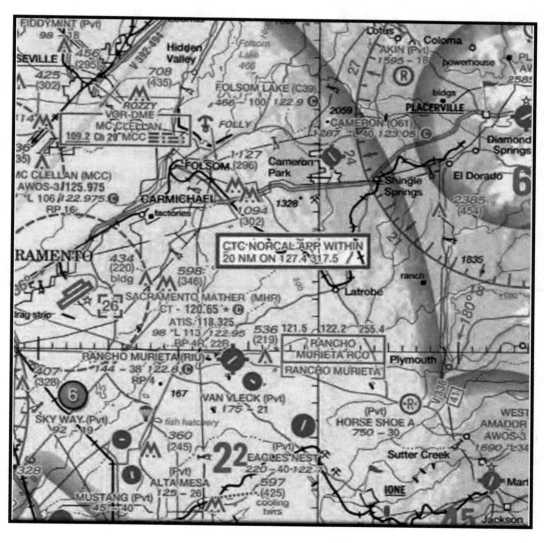

Figure 71, Southeast of Area 6. Not to scale.

Answer – c. According to the chart Legend Eagles Nest is a hard surfaced, 4,000' runway this is Private (see the Pvt above the name). Reference: Aero Chart, AIM

103. "RP" listed at the bottom of the airport information block for the Livermore Airport (KLVK), Figure 70 Area 5, stands for

a) Random Pattern (RP), meaning there are multiple traffic patterns in use.

b) Remote Pilot (RP), meaning there is frequent use of this airport by manned and unmanned aircraft.

c) Right Pattern (RP), meaning that the airport traffic pattern consists of right turns to the runway.

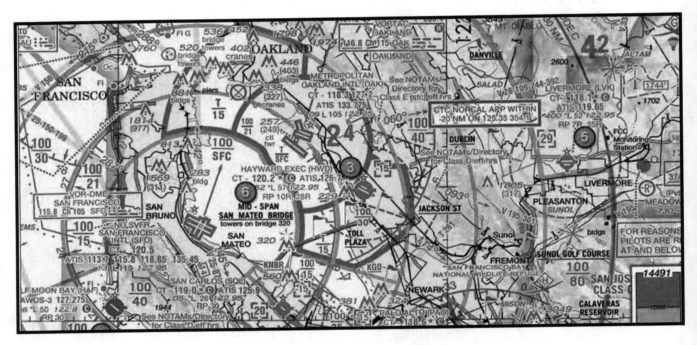

Figure 70, Area 5. Not to scale.

Answer – c. Use the Chart Legend to determine the definition of RP, which mean there is a Right (traffic) Pattern (RP) in use at this airport. The standard traffic pattern is Left and assumed unless otherwise stated. Since this airport has a control tower, aircraft arrivals will be routed by ATC vectors (or directions). However, there are times at which the tower is closed; this is when the RP will be utilized. Reference: Chart Legend

104. What is the definition and concerns regarding A-682, Figure 72, Area 9?

a) A-682 is an Action Area, no flight operations are allowed in this area under any circumstances except for Disaster Response and Search and Rescue (SAR) efforts. Under an exception, the Remote Pilot (RP) would need a Consent Letter (CS) signed by the commanding officer in charge of airspace operations.

b) A-682 is an Alert Area in which there is a high volume of military pilot training activities or an unusual type of aerial activity requiring extra vigilance on behalf of the UA flight crew.

c) A-682 is an (military) Action Area and requires permission from the controlling agency for flight operations.

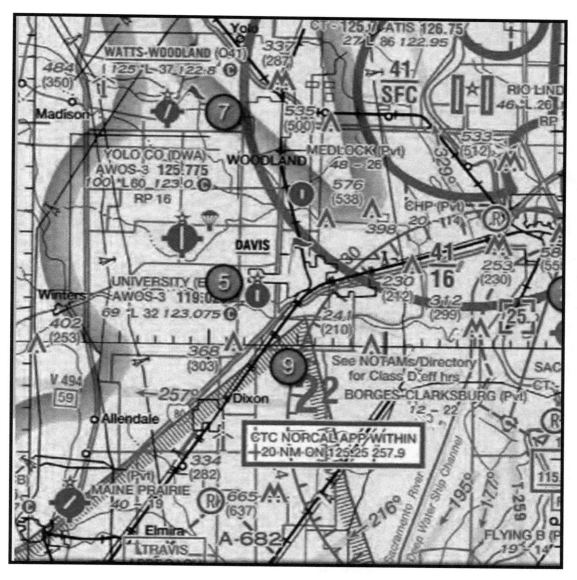

Figure 71, Area 9

Answer – a. Notice the magenta V-shaped hatched line that is over the Area 9 number, this is Alert Area A-682 (see number South of Area 9 near the bottom edge of the chart). Alert Areas are depicted similarly to a Military Operations Area (MOA), a close relative of Alert Areas. This is a type of military Special Use (SU) Airspace that is used for training and unusual types of aerial activities. Reference: Aero Chart, AIM

105. What are the vertical dimensions of the Class C Airspace over the Medlock Airport, Figure 71, Southeast of Area 7?

a) The Class C starts at the Surface (SFC) and extends to 4,100' MSL.

b) The Class C starts at 1,600' MSL and extends to 4,100' MSL.

c) Medlock Airport is not under the Class C Airspace referenced in the question.

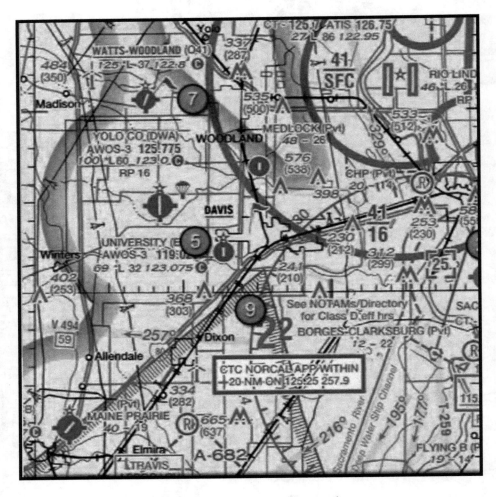

Figure 71, Area 7. Not to scale.

Answer – b. The magenta blue lines indicate Class C Airspace and the Medlock Airport is inside of the outer ring of the airspace (imagine an upside down wedding cake with the inner ring going to the surface). On the Southside of the outer ring are the figures 41/16, which means Class C begins at 1600' MSL and continues to 4100' MSL. Reference: Aero Chart, AIM

106. Figure 75, Area 1: A customer request aerial photos of a construction site that need to be completed the following day. The customer describes the mission area as being close to the Livermore Airport (KLVK) which is Class D airspace, Figure 74, Area 1. Without prior authorization from the Livermore ATC (via the FAA waiver form online), what is the minimum distance from KLVK Class D airspace acceptable to accept this mission (with standard Class D)?

a) Greater than 10 NM.

b) A safe distance outside of the current traffic pattern of the airport.

c) Greater than 5 SM (4.4 NM).

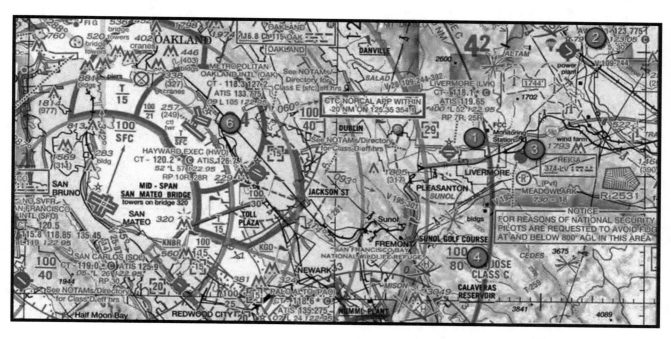

Figure 74, Area 1. Not to scale.

Answer – c. KLVK is the "classic size" of Class D airspace is 5 SM (4.4 NM) but can be modified to reflect the needs of the specific airspace (i.e. a keyhole cut-out for another airport). Since the customer wants the job done quickly, we deduce that permission has not been requested as a result of time constraints for gaining ATC approval on short notice. As a result, the flight would have to take place at a distance greater than 5 Statute Miles (SM) from KLVK. Caution: When a flight mission is in the immediate area outside of ATC controlled airspace is wise to pre-coordinate with ATC incase calculations are incorrect. Reference: Chart Legend, AIM, PHAK

107. What are the hours of the Lincoln Municipal Class D tower, Figure 53? What frequency would be used to monitor area traffic after hours?

a) Hours 1130 – 0600 Z. After hours monitor Lincoln Approach Control on 124.0 MHz.

b) Hours 0600 – 1130 Z. After hours monitor Unicom 122.95 MHz.

c) Hours 1130 – 0600 Z. After hours monitor Lincoln Tower on 118.5 MHz.

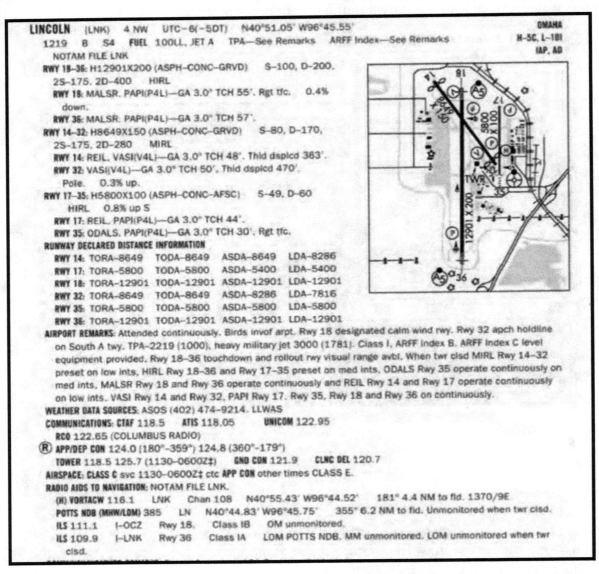

Figure 53, Chart Supplement (Lincoln Municipal Airport)

Answer - c. Look ¾ down the Chart Supplement page under "Airspace" and notice it states "SVC" (Service) 1130 – 0600 Z. Additionally noted is that the airspace is Class C during ATCT hours and converts to Class E after hours. When the tower is closed the Tower frequency is used, 118.5, which is also listed as the Common Traffic Advisory Frequency (CTAF). The Tower frequency does "double duty," so to speak, during and after ATCT hours. Tip: Use the Chart Supplement Legend to decode the page information. Reference: Chart Supplements

108.   You have been contracted to take aerial photos of a bridge inside of the Lincoln Class C Airspace. After submitting a waiver request to the FAA, you receive an email from the FAA stating that since the mission time is after hours, the airspace will revert to Class E. In this case, you will monitor the Tower frequency for aircraft arriving under a non-towered condition. The FAA representative suggests reviewing the traffic patterns about the job site. What are the traffic patterns in effect at Lincoln Municipal, Figure 53?

a) Right Runway 14 and Runway 32; to the left and on Runway 18 and Runway 35.

b) Right on Runways 14 – 32.

c) Left on Runway 14 and Runway 32; to the right on Runway 18 and Runway 35.

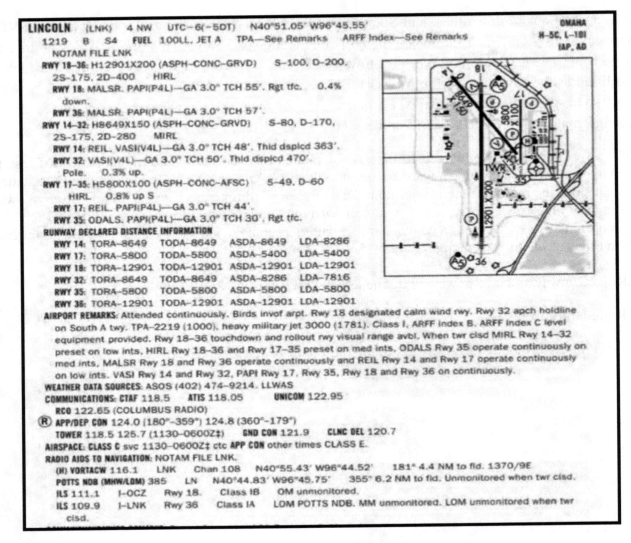

Figure 53, Chart Supplement (Lincoln Municipal Airport)

Answer – c. Definitely a tricky, and time-consuming question if you are not careful. Use the Chart Supplement Legend to decode the information. We know that all standard traffic patterns are to the left (left turns) into the airport. The Chart Supplement will state, any of the exceptions to the rule. Read just under the first few lines of text where it states RWY information. Looking down the list look for any reference to "Rgt tfc" (decoded Right Traffic). Runways 18 and 35 have Right Traffic (Rgt tfc) according to the text, all other runways are assumed to use standard left traffic. Reference: Chart Supplement

109. What Class of Airspace is the Lincoln Airport during the hours 1130 – 0600Z?

a) Class C

b) Class B

c) Class A

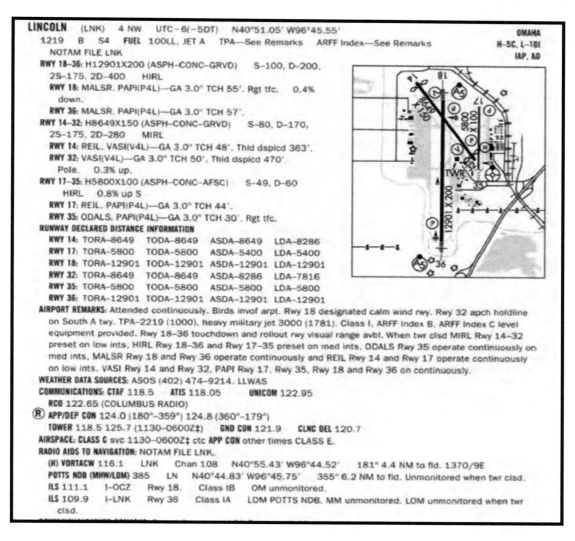

Figure 53, Chart Supplement

Answer - a. Look ¾ down the Chart Supplement page to "Airspace" and it states "Class C" and then gives information that Lincoln is Class E, outside of the ATCT hours of operation. Reference: Chart Supplement

110. How far and in what direction is the city of Lincoln from the airport, Figure 53?

a) 4 NM Northwest.

b) 6 NM South Direct.

c) 4 SM South.

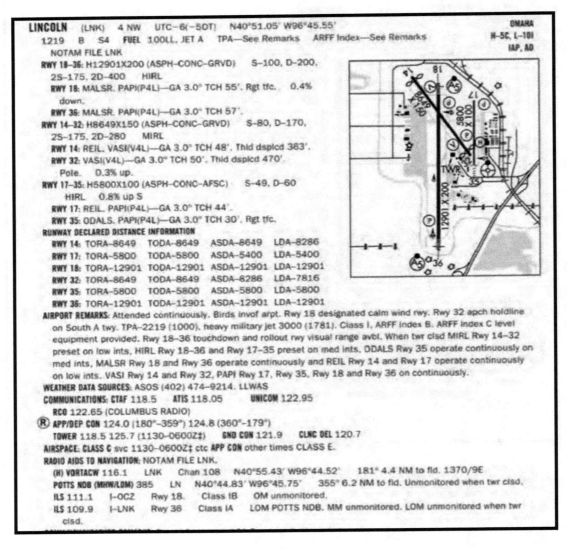

Figure 53, Chart Supplement

Answer – a. The answer is listed as the third piece of information on the Chart Supplement page, 4 NW, which means the city is 4 Nautical Miles (NM) to the Northwest of the airport. Use the Chart Supplement Legend to decode the information. Reference: Chart Supplement

111. A boat company wants to get some aerial video footage of their new boat for a marketing campaign, and the shoot will take place about 10 miles off the coast of North Carolina. Looking at the aeronautical chart, you notice this is considered a Warning Area. What is the significance of the Special Use Airspace called a Warning Areas that you typically find off the coast of the United States (US) such as in this example?

a) Warning Areas are exactly like Class A Airspace and should be treated as such by the UA flight crew.

b) They are similar in nature to restricted areas with regards to military activities but are less regulated. The United States government does not have sole jurisdiction over the airspace. A warning area is airspace of defined dimensions, extending from 3 NM outward from the coast of the US, containing activity that may be hazardous to nonparticipating aircraft.

c) They are dissimilar in nature to restricted areas; however, the United States government does have sole jurisdiction over the airspace. A warning area is airspace of defined dimensions, extending from 10 NM outward from the coast of the US, containing activity that may be hazardous to nonparticipating aircraft.

Answer – b. You would be in a Warning Area in this example. Warning Areas are typically found off the coast of the US and despite the official FAA definition above have been casually described as the "wild west" with military live firing and mission exercises.

112. Figure 80 and Figure 81: What is the traffic pattern for Runway 25 at the Crawford Airport?

a) Right-hand traffic pattern.

b) Left-hand traffic pattern.

c) It does not matter because it is an uncontrolled airport.

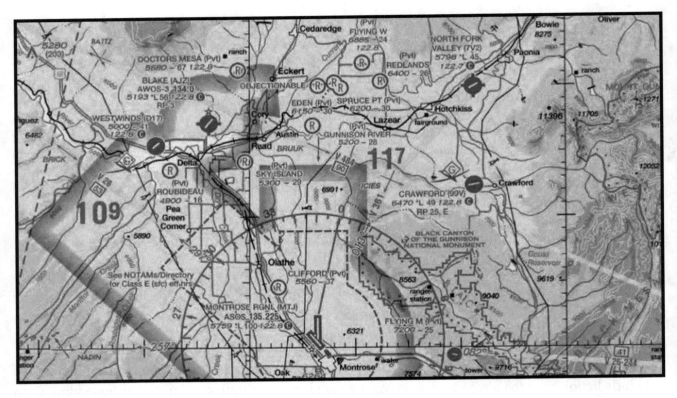

Figure 80. Not to scale.

216	**COLORADO**	

CRAWFORD (99V)   2 W   UTC −7(−6DT)   N38°42.25′ W107°38.62′   **DENVER** L−9E

6470   S2   OX 4   TPA—7470(1000)   NOTAM FILE DEN
RWY 07−25: H4900X20 (ASPH)   LIRL (NSTD)
  RWY 07: VASI (NSTD). Trees.   **RWY 25:** VASI (NSTD) Tank. Rgt tfc.
RWY E−W: 2500X125 (TURF)
  RWY E: Rgt tfc.   **RWY W:** Trees.
AIRPORT REMARKS: Attended continuously. Rwy 07−25 west 1300′ only 25′ wide. Heavy glider ops at arpt. Land to the
  east tkf to the west winds permitting. 100LL fuel avbl for emergency use only. Pedestrians, motor vehicles, deer
  and wildlife on and invof arpt. Unlimited vehicle use on arpt. Rwy West has +15′ building 170′ from thld 30′ left,
  +10′ road 100′ from thld centerline. +45′ tree 100′ L of Rwy 07 extended centerline 414′ from rwy end. −8′ to
  −20′ terrain off both sides of first 674′ of Rwy 25 end. E−W rwy occasionally has 6 inch diameter irrigation
  pipes crossing rwy width in various places. Rwy 07 has 20′ trees and −10′ to 20′ terrain 20′ right of rwy first
  150′. E−W rwy consists of +12 inch alfalfa vegetation during various times of the year. Arpt lgts opr
  dusk−0800Z‡. Rwy 07 1 box VASI left side for local operators only or PPR call 970−921−7700 or
  970−921−3018. Rwy 07−25 LIRL on N side from Rwy 25 end W 3800′. Rwy 07 1300′ from end E 300′. No thld
  lgts Rwy 07−25 3800′ usable for ngt ops.
COMMUNICATIONS: CTAF/UNICOM 122.8
RADIO AIDS TO NAVIGATION: NOTAM FILE MTJ.
  MONTROSE (H) VORW/DME 117.1   MTJ   Chan 118   N38°30.39′ W107°53.96′   033° 16.9 NM to fld. 5713/12E.

CREEDE
  MINERAL CO MEM   (C24)   2 E   UTC −7(−6DT)   N37°49.33′ W106°55.79′   **DENVER** H−3E, L−9E
  8680   NOTAM FILE DEN
  RWY 07−25: H6880X60 (ASPH)   S−12.5, D−70, 2D−110
    RWY 07: Thld dsplcd 188′.   **RWY 25:** Road.
  AIRPORT REMARKS: Unattended. Elk and deer on and invof arpt. Glider and hang glider activity on and in vicinity of
    arpt. Mountains in all directions. Departure to NE avoid over flight of trailers and resident homes, climb to 200′
    above ground level on centerline extended prior to turn. Acft stay to right of valley on apch and/or departure
    route. 2′ cable fence around apron.
  COMMUNICATIONS: CTAF 122.9
  RADIO AIDS TO NAVIGATION: NOTAM FILE DEN.
  BLUE MESA (H) VORW/DME 114.9   HBU   Chan 96   N38°27.13′ W107°02.39′   158° 38.1 NM to fld. 8730/14E.

Figure 81

Answer – a. The quickest way to determine the traffic pattern is to look on the aeronautical chart information bloc for Crawford Airport. On the last line of it states RP (Right Pattern) 25 (Runway 25). Reference: Aero Chart, Chart Supplement

113. Refer to Legend 15. What depicts a Class E airspace that begins at 700 feet AGL?

a) A dashed blue around an airport.

b) A dashed magenta around an airport.

c) A magenta vignette around an airport.

AIRSPACE

Information concerning Class B, C, and part-time D and E surface area airspace shall be published with effective times. Class D and E surface area airspace that is continuous as established by Rulemaking Docket will not be shown.
CLASS B—Radar Sequencing and Separation Service for all aircraft in CLASS B airspace.
CLASS C—Separation between IFR and VFR aircraft and sequencing of VFR arrivals to the primary airport.
TRSA—Radar Sequencing and Separation Service for participating VFR Aircraft within a Terminal Radar Service Area.
Class C, D, and E airspace described in this publication is that airspace usually consisting of a 5 NM radius core surface area that begins at the surface and extends upward to an altitude above the airport elevation (charted in MSL for Class C and Class D). Class E surface airspace normally extends from the surface up to but not including the overlying controlled airspace.
When part-time Class C or Class D airspace defaults to Class E, the core surface area becomes Class E. This will be formatted as:
AIRSPACE: CLASS C svc "times" ctc APP CON other times CLASS E:
or
AIRSPACE: CLASS D svc "times" other times CLASS E.
When a part-time Class C, Class D or Class E surface area defaults to Class G, the core surface area becomes Class G up to, but not including, the overlying controlled airspace. Normally, the overlying controlled airspace is Class E airspace beginning at either 700' or 1200' AGL and may be determined by consulting the relevant VFR Sectional or Terminal Area Charts. This will be formatted as:
AIRSPACE: CLASS C svc "times" ctc APP CON other times CLASS G, with CLASS E 700' (or 1200') AGL & abv.
or
AIRSPACE: CLASS D svc "times" other times CLASS G with CLASS E 700' (or 1200') AGL & abv:
or
AIRSPACE: CLASS E svc "times" other times CLASS G with CLASS E 700' (or 1200') AGL & abv.
NOTE: AIRSPACE SVC "TIMES" INCLUDE ALL ASSOCIATED ARRIVAL EXTENSIONS. Surface area arrival extensions for instrument approach procedures become part of the primary core surface area. These extensions may be either Class D or Class E airspace and are effective concurrent with the times of the primary core surface area. For example, when a part-time Class C, Class D or Class E surface area defaults to Class G, the associated arrival extensions will default to Class G at the same time. When a part-time Class C or Class D surface area defaults to Class E, the arrival extensions will remain in effect as Class E airspace.
NOTE: CLASS E AIRSPACE EXTENDING UPWARD FROM 700 FEET OR MORE ABOVE THE SURFACE, DESIGNATED IN CONJUNCTION WITH AN AIRPORT WITH AN APPROVED INSTRUMENT PROCEDURE.
Class E 700' AGL (shown as magenta vignette on sectional charts) and 1200' AGL (blue vignette) areas are designated when necessary to provide controlled airspace for transitioning to/from the terminal and enroute environments. Unless otherwise specified, these 700'/1200' AGL Class E airspace areas remain in effect continuously, regardless of airport operating hours or surface area status. These transition areas should not be confused with surface areas or arrival extensions.
(See Chapter 3, AIRSPACE, in the Aeronautical Information Manual for further details)

Legend 15

Answer – c. You don't necessarily need the Airport Supplement to answer this question. Never-the-less, Legend 15 describes the Class E Airspace towards the bottom of the page. In the last paragraph states, "Class E 700' AGL (shown as a magenta vignette on a sectional chart). A magenta vignette, a.k.a. magenta circle. However, you could alternatively look at the graph legend of the chart. Reference: Chart Supplement Legend

114. What information is contained in the Notices to Airmen Publication (NTAP)?

a) Current NOTAM (D) and FDC NOTAMs.

b) All current NOTAMs only.

c) Current Airport Facility Directory, information and FDC NOTAMs.

Answer – a. Tip: Since NOTAM is mentioned in the question, it is also listed in the Answer. Since there are two answers with NOTAMs mentioned you need to consider FDC NOTAMs, but first let us review D NOTAMs. NOTAMs are information that is not shown on current charts, instrument approach plates (for pilots flying instrument approaches to an airport), etc. D NOTAMs consist of information regarding closures/construction at the airport. For the UA pilot it also includes uncharted obstructions, service notifications (like tower closure). Additionally, event information such as airshows and fly-ins are considered D NOTAMs. The Flight Data Center (FDC) NOTAMs cover changes not reflected on charts related to airspace, instrument approaches (changes to the approach itself, not just obstructions and closures like with D NOTAMs) and airways. Reference: AIM

115. When NOTAMs are published in the Notices to Airmen Publication (NTAP) they are

a) still part of a standard weather briefing.

b) only available in a standard weather briefing if specifically requested.

c) canceled and are no longer valid.

Answer – b. A weather briefer many include some NOTAMs with the briefing, but not Flight Data Center (FDC) NOTAMs unless specifically requested. FDC NOTAMs cover changes not reflected on charts related to airspace, instrument approaches (changes to the approach itself, not just obstructions and closures like with D NOTAMs) and airways. If not given by the briefer you should also ask for D NOTAMs if not given. We suggest asking the briefer, "are there any NOTAMs, FDC or D NOTAMs, that will affect the area of my mission?" Some suggest that using the online briefing as an easier way to scan the data for items that may affect a flight operation due to the large quantities of NTAP data. Reference: AIM

116. The angular difference between true north and magnetic north is referred to as

a) magnetic deviation.

b) magnetic variation.

c) compass acceleration error.

Answer – b. Magnetic Variation is listed on the aeronautical chart as long magenta dashed lines that travel semi-vertical on the chart (see below) that used to illustrate the conversion factor of true north from magnetic north. This is used by manned aircraft pilots in determining the course of cross-country flights.

117. You have a flight mission in the town of Guyton, Southwest of Area 3, Figure 23 (near the Briggs Airport. As a safety precaution, you decide to fill out the FAA airspace waiver form with regards to the Savannah ATC. Describe the location of Guyton giving radial and distance from the Savannah VOR. What is, the radial and distance from the SAV VOR?

a) 10 NM on the 120° radial.

b) 16 NM on the 325° radial.

c) 19 NM on the 280° radial.

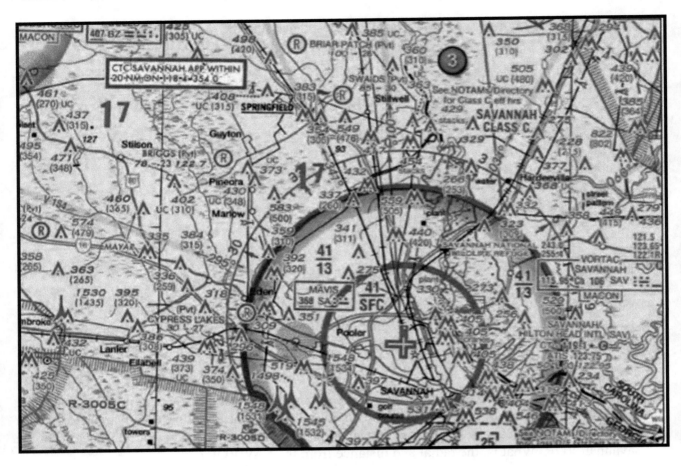

Figure 23. Not to scale.

Answer – b. It's best to use a chart plotter on a hard copy of this figure until you have done several computations. Noticed the compass rose that is shown slightly outside of the Savannah Class C Airspace, this is for the VOR navigation facility located on the airport (notice there is no VOR symbol on the chart). For a rough estimate you can see, there is a Victor Airway crossing the compass rose, indicating 341° radial (think of a line radiating from the center, like spoke). Just to the left of the 341° radial, you can barely see the 330° radial because it is partially hidden under the numbers 1-7 (more on this later). Use your plotter and draw a line from the center of the VOR, and/or airport, to the town of Guyton. Where this line crosses the compass rose is part of your answer, which looks like the 325° radial to me.

The remaining item to calculate is the NM from the VOR. You can use the plotter to determine the 16 NM distance and crosscheck by knowing the Savannah Class C is of a standard design and extends out 10 NM from the airport (5 NM to the first ring then another 5 to the outer ring). From this, you can roughly estimate the distance is 15 NM, which is very close to the 16 NM answer. Reference: Aero Chart, PHAK

118. Figure 23, Area 3. You have a flight mission adjacent to the Ridgeland Airport (3J1), northeast of Area 3. What potential safety issue(s) can you determine by a first glance at the aeronautical chart? Determine the radial and distance from the Savannah (SAV) VOR you can relay to those responsible for area airspace.

a) The mission is adjacent to the Ridgeland Airport (3J1), 24 NM on the 033° radial of the SAV VOR. It is required to notify the 3J1 airport owner/operator regarding the flight mission. Notice the glider operations (see the glider chart symbol) at the 3J1 airport, which means glider towing and related activity could be a factor. Also, the airport is designated as a checkpoint for aircraft approaching Savannah Class C. 3J1 also rest under the Beaufort 3 Military Operations Area (MOA), and the edges of the aeronautical chart should be checked for the details of altitudes and times of coverage.

b) The mission is adjacent to the Ridgeland Airport (3J1), 19 NM on the 330° radial of the SAV VOR. No safety issues are present besides due diligence of a visual inspection of the mission area once on-site.

c) The mission is adjacent to the Ridgeland Airport (3J1), 17 NM on the 133° radial of the SAV VOR. It is required to notify the 3J1 airport owner/operator regarding the flight mission. Notice the glider operations (see the glider chart symbol) at the 3J1 airport, which means glider towing and related activity could be a factor. Also, the airport is designated (with a flag on the chart) as a checkpoint for aircraft approaching Savannah Class C. 3J1 also rest under the Beaufort 3 Military Operations Area (MOA), and the edges of the aeronautical chart should be checked for the details of altitudes and times of coverage.

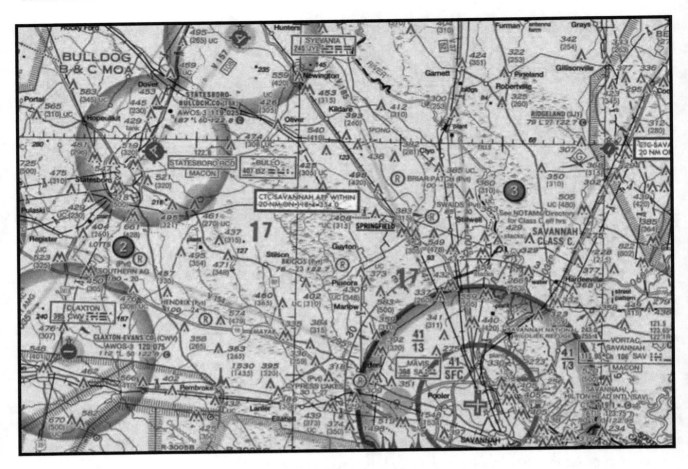

Figure 23, Area 3. Not to scale.

Answer – a. First, find the mileage and radial 3J1 is from the SAV VOR. The VOR is located on the Savannah Airport, place your plotter on the airport and draw a line to the 3J1 Airport to determine the mileage which is 24 NM. Determining the radial can be done using the compass rose that can be seen superimposed just past the SAV Class C outer ring. Calculating the radial is easier than normal because there is a Victor Airway with a radial of 034° that passes over the airport. 034° is slightly off from the center of the airport, so the best answer is 033°.

"First glance" potential safety issues could be; glider operations at 3J1 (see the glider symbol), the airport is listed as a checkpoint (see the small flag symbol) for arrivals to Savannah (for additional crossing traffic), the Military Operations Area (MOA) that surrounds 3J1 (see magenta hatched line), and the Victor Airway that crosses the airport (additional aircraft traffic). Reference: Aero Chart, Chart Legend, PHAK, AIM

119. Figure 24, the southwest area of Savannah (SAV) Class C: You have a flight mission within the cluster of 3 large antennas 9 miles Southwest of the SAV airport (no altitude waiver). Who should you contact for permission to fly in this area? Determine the radial and distance from the Savannah (SAV) VOR. Are there any other safety concerns in the area you can see at first glance?

a) You should contact the closest airport owner/operator, Cypress Lakes private airport for permission. Traffic pattern operations at Cypress Lakes should be considered on this mission. The mission area is 110° radial at 12 NM.

b) You must send a certified letter to the Savannah (SAV) ATC facility notifying them of the date, time, altitude and location of the mission, no further action or response necessary. The mission area should be described as 15 NM from the center of the city of Savannah and not as a radial from the VOR. Safety concerns will consist of numerous glider tow operations in the area.

c) Permission is not mandatory while flying under the shelf of the Savannah (SAV) Class C which starts at 1,300' MSL. The cluster of 3 antennas are 9 NM Southwest of the SAV VOR on the 240° radial. Safety concerns: In addition to the guy wires of the antennas there is also a Restricted area just to the south of the mission site.

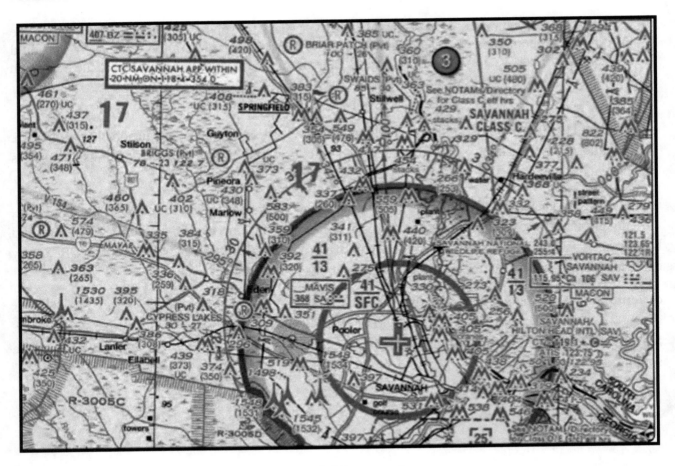

Figure 23. Not to scale.

Answer – c. It is not required to get permission from SAV Class C -ATC since the shelf of Class C begins at 1,300' MSL. However, it is wise to contact the SAV ATCT to coordinate.

The mission site is just to the inside of the Compass Rose of the SAV VOR at 9 NM on the 240° radial (notice the slight offset between the SAV VOR Compass Rose and the SAV Class C). You can use a plotter to determine the mileage (by drawing a line from the airport VOR to the center of the antennas) or remember the standard Class C extends to a 10 NM radius and use that as a reference. Calculating the radial can be done with a plotter or done by eye. Notice the number 24 close to the R-3005 (Restricted Airspace), 24 is the 240° radial reference from the VOR. Using your plotted line from the airport VOR you can see it passes through the 240° radial.

As a safety precaution due to the close proximity of Restricted Airspace (R-3005), we suggest contacting the controlling agency.

Lastly, be sure to consider the guy wires associated with the antennas, or "wires environment" as the FAA likes to call it. Guy wires can extend 2,000' from the primary structure.

120. What are the minimum flight visibility and cloud clearances for sUAS operators?

a) 1 SM average slant range visibility, 2,000' below clouds and 500' horizontally from the cloud.

b) 1 SM and clear of clouds.

c) 3 SM average slant range visibility, 500' below clouds and 2,000' horizontally from the cloud.

Answer – c. Memory aid: 3, 5 and 2 (3 SM, 500' below, 2,000' from). Reference: CFR 107.51

121. A sUAS aircraft cannot fly higher than 400' AGL (no waiver) or

a) flown within a 200' radius above a structure's immediate uppermost limit.

b) flown within a 4,000 horizontal radius of the Control Station (CS).

c) flown within a 400' radius above a structure's immediate uppermost limit.

Answer – c. The 400' AGL altitude limit extends to a structure. Reference: CFR 107.51

Aviation Weather

122. Every physical process of weather is accompanied by or is the result of a

a) movement of air.

b) pressure differential.

c) heat exchange.

Answer – c. By-in-large weather is created due to the unequal heating of the Earth's surface. On the macro scale, bodies of water are heated by the Sun differently than the ice in the Arctic. The land mass is heated differently than the Ocean. On a micro scale, a plowed field is different than a field of green grass or large areas of rock. The end result is the weather. Reference: AWH

123. Convective circulation patterns associated with sea breezes are caused by

a) warm, dense air moving inland from over the water.

b) water absorbing and radiating heat faster than the land.

c) cool, dense air moving inland from over the water.

Answer – c. The term convective circulation is a very accurate description because the air traveling over cool water (cooling the air) moves inland and is then heated by the land traveling up and back out to sea to start the process again in a circular pattern. Reference: AWH

124. The term used to describe the boundary between two different air masses is a

a) frontolysis.

b) frontogenisis.

c) front.

Answer – c. The dividing line between two different air masses is known as a front. There are four major types of fronts; cold, warm, stationary and occluded. Reference: AWH

125. What happens to the winds when a front passes through a mission flight area?

a) A change in wind direction.

b) The shift in the type of precipitation.

c) Radical instability of the air mass.

Answer – a. A change in wind direction will signify the passing of a front. Reference: AWH

126. A major characteristic when a front is crossing the mission flight area is

a) a change in the temperature.

b) an increase in cloud coverage.

c) an increase in relative humidity.

Answer – a. Fronts are areas, of weather characteristically similar. When crossing frontal areas there will be a change in wind direction and temperature. Reference: AWH

127. What weather phenomenon can be expected with thunderstorm activity in the vicinity of a flight mission area?

a) Precipitation static.

b) Wind-shear turbulence.

c) Steady rain.

Answer – b. Thunderstorms, aside from the obvious effects, contain vertical up and downdrafts within that can be more dangerous with size. The width of the thunderstorm can have a significant effect, but the vertical development creates drafts necessary for wind shear. The consistent change in wind direction is defined as wind-shear and bring about associated turbulence. The effect of wind-shear can be experienced, well away from the thunderstorm (as much as 20 NM depending on its size and Level of intensity). Reference: AWH

128. What is the narrow band of active thunderstorms called that can develop ahead of a cold front (not part of the front)?

a) pre-frontal system.

b) squall line.

c) dry line.

Answer – b. A squall line is a dangerous line of storms that can develop ahead of a cold front. Reference: AWH

129.   What conditions are necessary for the formation of thunderstorms?

a) High humidity, lifting force, and unstable conditions.

b) High humidity, high temperature, and cumulus clouds.

c) Lifting force, moist air, and extensive cloud cover.

Answer a. Some of the other answers sound very reasonable, but the answer the FAA is looking for is high humidity, lifting force, and unstable conditions. Instability is a must, and humidity provides the fuel. When combined with a lifting force (for example, topographic lifting, i.e. mountains) gives lift to the air mass, a perfect recipe for a thunderstorm. Reference: AWH

130.   During the life cycle of thunderstorms, which stage is characterized predominately by downdrafts?

a) Cumulus.

b) Dissipating.

c) Mature.

Answer – b. The life cycle of thunderstorms follows from; cumulus, mature to dissipation. The Cumulus stage is the large white "puffy clouds." Water droplets start to form and grow as the up and down drafts intensify. As its strength builds, with a lot of up and down draft movement/lifting it becomes Mature. The thunderstorm reaches a point in which the storm cannot hold the water drop, signifying the Dissipating (as in dissipating the water) stage. Reference: AWH

131.   Thunderstorms reach their greatest intensity during the

a) mature stage.

b) downdraft stage.

c) cumulus stage.

Answer – a. The stages of a thunderstorm are; Cumulus, Mature, and Dissipating. In the Mature stage, there are up and down drafts of building water droplets in the storm-cloud. This is the largest and most powerful stage of the thunderstorm just before Dissipating (as in dissipating the water). Reference: AWH

132. What feature is normally associated with the Cumulus stage of a thunderstorm?

a) Roll cloud.

b) Continuous updraft.

c) Frequent lightening.

Answer – b. The Cumulus ("puffy white cloud") stage of a thunderstorm is known for the building of the cloud via updrafts. This will eventually lead to the powerful Mature stage of the storm. Reference: AWH

133. Which weather phenomenon signals the beginning of the mature stage of a thunderstorm?

a) The appearance of an anvil top.

b) Precipitation beginning to fall.

c) Maximum growth rate of the clouds.

Answer – b. The Mature stage of a thunderstorm is not the Dissipating (as in dissipating water) but is still known for precipitation. Reference: AWH

134. Thunderstorms that generally produce the most intense hazard to manned aircraft in-flight are

a) squall line thunderstorms.

b) steady-state thunderstorms.

c) warm front thunderstorms.

Answer – a. All thunderstorms are bad for sUAS flight operations, but the FAA may ask about hazards specific to manned aircraft for an overall understanding of the topic. Squall lines are the thunderstorms that form ahead of cold fronts and are known to be particularly dangerous. Reference: AWH

135. Which weather phenomenon is always associated with a thunderstorm?

a) Lightening.

b) Heavy rain.

c) Hail.

Answer – a. You may get hail and heavy rain from a thunderstorm but lightning will always be a defining characteristic associated with thunderstorms. Reference: AWH

136. The mature stage of a thunderstorm, begins with

a) formation of the anvil tops.

b) the start of precipitation.

c) continuous downdrafts.

Answer – b. The Mature stage of a thunderstorm is not the Dissipating (as in dissipating water) but is still known for precipitation. Reference: AWH

137. What is the space between thunderstorms called?

a) Trough

b) Front

c) Wave

Answer- b. This is a difficult question, but the answer isn't (a) trough, I'm leaning towards a wave. In other questions, they discuss the fact that thunderstorms form ahead of fronts. Since the question says, in-between and not ahead, front, we don't believe this is the correct answer. If a thunderstorm is an area of low pressure then outside of it is higher pressure, hence a wave.

138. In what phase of a thunderstorm does the most turbulence occur?

a) Cumulus

b) Mature

c) Dissipating

Answer – c. As the thunderstorm is dissipating all of its water intense downdrafts occur, causing the greatest turbulence. Reference: PHAK, AWH

139. What is the best type of conditions for a sUAS flight mission?

a) High lift days with rotor and lenticular clouds.

b) Cool low humidity days with a stratiform cloud layer.

c) Directly under virga.

Answer – c. The answer will contain stable atmospheric traits such as low humidity and stratiform clouds. Rotor and lenticular clouds are indicators of great turbulence. Virga, streams of clouds and moisture under the base of a cloud indicate turbulence. Reference: PHAK, AWH

140. One in-flight condition necessary for structural icing to form

a) small temperature/dew point spread.

b) Stratiform clouds.

c) Visible moisture.

Answer – c. Structural icing can seriously affect the flight of a sUAS with an increase in weight and the lack of smooth airflow over lifting surfaces (wings and propellers). Ice requires water or visible moisture as the FAA says, to form.

Clouds are also considered visible moisture, but it is not specific to a cloud type, such as Stratiform clouds. A small temperature – dew point spread refers to the conditions conducive for precipitation as the temperature and dew point meet. When temperature and dew point meet, it is the temperature at which the air can't hold any additional moisture and some type of precipitation will form. Reference: AWH

141. In which environment is aircraft structural ice most likely to have the highest accumulation rate?

a) Cumulus clouds with below freezing temperatures.

b) Freezing drizzle.

c) Freezing rain.

Answer – c. As you might imagine drizzle would be less likely, then rain, to produce heavy structural icing accumulation rates. Cumulus clouds cannot produce the structural icing that freezing rain. Reference: AWH

142. The presence of ice pellets at the surface is evidence that there

a) are thunderstorms in the area.

b) has been a cold frontal passage.

c) is a temperature inversion with freezing rain at a higher altitude.

Answer – c. A temperature inversion means that the air temperature rises with altitude verses lowering as altitude increases (think of driving up a mountain, going up is associated with lower temperature). Ice pellets, a form of precipitation falling through a large layer of air that is below freezing allowing the formation of pellets. Reference: AWH

143.   A lens-shaped cloud that typically forms over mountainous areas contain dangerous winds are called

a) an inactive frontal cloud.

b) a funnel cloud.

c) a lenticular cloud.

Answer – c. Mountainous areas are known best for lenticular clouds, lens-shaped clouds that form over mountain tops that are synonymous with dangerous winds and turbulence. Reference: AWH

144.   Stationary lens-shaped clouds that form over mountain peaks and ridges that are associated with mountain waves are known as

a) Cumulus clouds.

b) Standing lenticular clouds.

c) Vortex clouds.

Answer – b. Lenticular clouds are lens-shaped clouds that contain dangerous winds and turbulence. Since they are stationary over a mountain top or ridge, they are called "Standing" but don't let the docile shape trick you. Reference: AWH

145.   Mountain waves and their related turbulence can be anticipated when winds 40 knots or greater, and the wind proceeds

a) across a mountain ridge containing stable air.

b) down a mountain valley containing unstable air.

c) parallel to a mountain peak containing stable air.

Answer – a. Stable, high winds, 40 knots is greater going over a mountain ridge, are the required components for mountain wave turbulence. Reference: AWH

146.   Where does wind shear occur?

a) Only at higher altitudes.

b) Only at lower altitudes.

c) At all altitudes, in all directions.

Answer – c. Wind Shear is associated with a location downstream from thunderstorms creating areas of rapidly shifting winds. As a result, pilots can experience turbulence and rapid fluctuations in airspeed. Manned aircraft pilots will report fluctuations in airspeed to ATC, for example, "we are experiencing airspeed fluctuation plus and minus 10 knots." As stated in the answer Wind Shear can happen in any altitude. Reference: AWH

147.   When may hazardous wind shear be expected?

a) When stable air crosses a mountain barrier where it tends to flow in layers forming lenticular clouds.

b) In areas of low-level temperature inversion, frontal zones, and clear air turbulence.

c) Following frontal passage when stratocumulus clouds form indicating mechanical mixing.

Answer – b. Frontal activity, the crossing of weather systems can spur adverse weather conducive to wind shear. Additionally, low-level temperature inversions and Clear Air Turbulence (CAT), a close cousin to wind shear, can also be conducive to such conditions. Reference: AWH

148.   What is meant by the term "dew point?"

a) The temperature at which condensation and evaporation are equal.

b) The temperature at which dew will always form.

c) The temperature to which air must be cooled to become saturated.

Answer – c. The air has a saturation (or maximum) point in which it can hold moisture (or water). Once saturation occurs some form of precipitation will occur. This point of saturation is represented by temperature, called the dew point temperature. When the outside air temperature reduces to the dew point, some form of precipitation will be present. Reference: AWH

149. The amount of water vapor, which air can hold depends on the

a) dew point.

b) air temperature.

c) stability of the air.

Answer – b. As the temperature is increased the air can hold more water vapor. Memory aid: As temperature increases the air becomes thinner, or the molecules of air are more spread out, making way for more capacity to hold water (vapor). Reference: AWH

150. What are the processes by which moisture is added to unsaturated air?

a) Evaporation and sublimation.

b) Heating and condensation.

c) Supersaturation and evaporation.

Answer – a. Most people are familiar with evaporation, for example, lake water evaporating into the atmosphere. The creation of snow and ice are an example of sublimation, where water vapor converts directly to a solid, bypassing the solid (water) phase. Reference: AWH

151.   Which conditions result in the formation of frost?

a) The temperature of the collecting surface is at or below freezing when small droplets of moisture fall on the surface.

b) The temperature of the collecting surface is at or below the dew point of the adjacent air, and the dew point is below freezing.

c) The temperature of the surrounding air is at below freezing when small drops of moisture fall on the collecting surface.

Answer – b. The conditions for frost are that the collecting surface is at or below the dew point and that the dew point adjacent air is below freezing. In a nutshell, collecting surface at or below a dew point temperature this is below freezing.

152.   Clouds, fog, or dew will always form when

a) water vapor condenses.

b) water vapor is present.

c) relative humidity reaches 100 percent.

Answer – a. These are all examples of water vapor becoming fog (essentially a cloud at ground level), clouds, or dew as long as the correct condensation nuclei, is present. Condensation nuclei is a small particle, such as dust, with which vapor can condense into clouds, fog or dew. Reference: AWH

153.   What type of fog can result in low-level turbulence and dangerous icing conditions?

a) Rain-induced fog.

b) Upslope fog.

c) Steam fog.

Answer – c. Steam fog forms when cold air moves over warmer water, which can cause icing and low-level turbulence. Reference: AWH

154. In which situation is advection fog most likely to form?

a) A warm, moist air mass on the windward side of mountains.

b) An air mass moving inland from the coast in winter.

c) A light breeze blowing colder air out to sea.

Answer – b. Advection (movement of air) fog happens when moist air moves over a cool surface. Reference: AWH

155. What situation is most conducive to the formation of radiation fog?

a) Warm, moist air over low, flatland areas on clear, calm nights.

b) Moist, tropical air moving over cold, offshore water.

c) The movement of cold air over much warmer water.

Answer – c. Radiation fog occurs when the warmer water (or land) heats the cold air above its saturation point producing fog. Reference: AWH

156. What types of fog depend upon the wind in order to exist?

a) Radiation fog and ice fog.

b) Steam fog and ground fog.

c) Advection flog and upslope fog.

Answer – c. The term "advection" means movement and upslope fog is the movement of air upwards or upslope. Reference: AWH

157. Clouds are divided into four families, according to;

a) outward shape.

b) height range.

c) composition.

Answer – b. Think of cumulous, tall billowy clouds, being different than thin stratus clouds. These two examples are very different in height. Reference: AWH

158. The suffix "nimbus," used in naming clouds means

a) a cloud with extensive vertical development.

b) a rain cloud.

c) a middle cloud containing ice pellets.

Answer – c. Nimbus are clouds in the 6,500' – 20,000' AGL range, which are considered middle clouds. Reference: AWH

159. The conditions necessary for the formation of cumulonimbus clouds are lifting action and

a) unstable air containing an excess of condensation nuclei.

b) unstable, moist air.

c) either stable or unstable air.

Answer – b. Cumulonimbus, are middle clouds (hence the use of nimbus on the end of cumulous) which range from 6,500' – 20,000' in size. These clouds have "extensive vertical development," in FAA-speak, but tall for most of us, which requires moist, unstable air to create. Reference: AWH

160.   What clouds have the greatest turbulence?

a) Towering cumulus.

b) Cumulonimbus.

c) Nimbostratus.

Answer – b. Cumulonimbus, are middle clouds (hence the use of nimbus on the end of cumulous) which range from 6,500' – 20,000' in size. These clouds have "extensive vertical development," in FAA-speak, but taller than most of us, which means extensive up and down draft, or turbulence. Towering cumulus, are found in the low or middle height ranges. Reference: AWH

161.   What cloud types would indicate convective turbulence?

a) Cirrus clouds.

b) Nimbostratus clouds.

c) Towering cumulus clouds.

Answer – c. Cirrus and Stratus clouds are relatively thin as compared to the Towering Cumulus clouds. Towering Cumulus have extensive vertical development (hence towering) containing many up and down drafts which mean turbulence. Additionally, being convective (creating thunderstorms) in nature also indicates turbulence. Reference: AWH

162.   What is a characteristic of stable air?

a) Stratiform clouds.

b) Unlimited visibility.

c) Cumulus clouds.

Answer – a. Stratiform (Stratus) clouds are thin with little vertical development (height). As a result, there are little (to no) up and down drafts like you would find in Cumulus clouds. Stratus clouds mean stable air conditions. Reference: AWH

163. When warm, moist, stable air flows upslope it

a) produces stratus type clouds.

b) causes showers and thunderstorms.

c) develops convective turbulence.

Answer – a. The key in this question is stable air, which means the relatively thin stratus type cloud. If you take a warm, moist and stable air mass upslope on a mountain, the temperature decreases to the dew point producing visible moisture, i.e. stratus type clouds. Reference: AWH

164. If an unstable air mass is forced upward what type of clouds can be expected?

a) Stratus clouds with little vertical development.

b) Stratus clouds with considerable associated turbulence.

c) Clouds with considerable vertical development and associated turbulence.

Answer – c. The question describes "unstable" air, meaning clouds that produce extensive vertical development (height), cumulonimbus for example. Reference: AWH

165. What are the characteristics of unstable air?

a) Turbulence and good surface visibility.

b) Turbulence and poor surface visibility.

c) Nimbostratus clouds and good surface visibility.

Answer – a. The word "unstable" can be a little misleading. Instead, think of the big, billowy clear weather Cumulous clouds. Unstable air describes cloud types that contain extensive vertical development, Cumulous clouds. These clouds contain up and downdrafts as they build, but the visibility on the surface is good. Reference: AWH

166. A stable air mass is most likely to have which characteristic?

a) Showery precipitation.

b) Turbulent air.

c) Poor surface visibility.

Answer – c. Stable air masses are required for Stratus type clouds, those with little vertical development (height) and thus little up and down drafts, meaning stable air. Think of those gray Winter days that have Stratus clouds, smooth flying but poor surface visibility. If visibility is good enough, the winds and light make for good aerial photography flying weather. Reference: AWH

167. Steady precipitation preceding a front is an indication of

a) Stratiform clouds with moderate turbulence.

b) Cumuliform clouds with little or no turbulence.

c) Stratiform clouds with little or no turbulence.

Answer – c. The term "steady precipitation" is a tip-off because clouds with little vertical development (height), like Stratiform (Stratus), produce more mild effects such as precipitation and turbulence. Reference: AWH

168. What are the characteristics of a moist, unstable air mass?

a) Cumuliform clouds with showery precipitation.

b) Poor visibility and smooth air.

c) Stratiform clouds and showery precipitation.

Answer – a. The words "unstable air mass" means clouds with extensive vertical development (height), such as Cumuliform (Cumulous) clouds. When Cumulous clouds reach the Mature stage they can produce showery precipitation. Reference: AWH

169.   What measurement can be used to determine the stability of the atmosphere?

a) Atmospheric pressure.

b) Actual lapse rate.

c) Surface temperature.

Answer – b. The actual lapse rate is based on the fact that stable air masses typically decrease in temperature as altitude increases (think of driving up a mountain). Based on this idea, a measure of atmospheric stability can be derived from this definition. Reference: AWH

170.   What would decrease the stability of an air mass?

a) Warming from below.

b) Cooling from below.

c) Decrease in water vapor.

Answer – a. Warming an air mass from below promotes vertical development, an unstable characteristic. Reference: AWH

171.   What feature is associated with a temperature inversion?

a) A stable layer of air.

b) An unstable layer of air.

c) Chinook winds on mountain slopes.

Answer – a. Typically, air temperature decreases with altitude (think of driving up a mountain). A temperature inversion is an opposite effect, temperature rising with an increase in altitude. Instead of having any possible warming from below for vertical cloud development the air produces a stable condition, perfect for Stratiform clouds. Reference: AWH

172. The most frequent type of ground or surface-based temperature inversion is that which is produced by

a) terrestrial radiation on a clear, relatively still night.

b) warm air being lifted rapidly aloft in the vicinity of mountainous terrain.

c) the movement of colder air under warm air, or the movement of warm air over cold air.

Answer – a. A great example of ground or surfaced based inversion is the heating of the Earth (terrestrial) by the Sun during the Winter. Since the Sun is low on the horizon, it heats the Earth and as night-time approaches the heated air rises to a low level (like a lid) with a cooler mass of air directly to the surface. The consequence is that pollution (home fires, industrial, etc.) can suppress in this lower, colder lay of surface air. Reference: AWH

173. When there is a temperature inversion you would expect to experience

a) clouds with extensive vertical development above and inversion aloft.

b) good visibility in the lower levels of the atmosphere and poor visibility above an inversion aloft.

c) an increase in temperature as altitude increases.

Answer – c. In this answer simply keep in mind the core definition of a temperature inversion, an increase in temperature as altitude increases. Answer a is incorrect because inversions are not conducive to extensive vertical development as when heating is done from below. Answer b is incorrect because the visibility would be poor at the surface because the warmer air above traps the air, along with the pollution, at the surface. Reference: AWH

174. Which weather conditions should be expected beneath a low-level temperature inversion layer when the relative humidity is high?

a) Smooth air, poor visibility, fog, haze, or low clouds.

b) Light wind shear, poor visibility, haze, and light rain.

c) Turbulent air, poor visibility, fog, low stratus type clouds, and showery precipitation.

Answer – a. Smooth, stable air is not conducive to extensive vertical cloud development. The warm air above the inversion acts as a lid on the air below producing, poor visibility, fog, haze or low clouds. Reference: AWH

Automated Flight Service Station (AFSS) Weather Services

175. Zulu (Z), a.k.a. Greenwich Mean Time (GMT), is used for

a) Time limited parts, calculations.

b) Aviation weather reports.

c) Not used, only Eastern Standard Time is used.

Answer – b. Aviation is on a standard clock, the GMT or Zulu (Z) time to convey timely weather information (in addition to world-wide aircraft scheduling) to pilots. As the saying goes, "the day starts in Greenwich, England," and all other time zones have a conversion to aid pilots in converting from local time to GMT. Most areas observe "Daylight Saving Time" which means the conversion factor will change slightly as you "Spring forward" or "Fall back." The Reference: AWH

176. When calculating Zulu or Greenwich Mean Time (GMT) the local time of the flight mission is converted based on

a) Sexton calculations.

b) a 24-hour clock. For example; 0700 is 7 a.m. and 1700 is 5 p.m. Local time is then converted to Zulu or GMT with the appropriate conversion for the time zone the flight mission is taking place.

c) a 12-hour clock. For example; 0700 is 7 a.m. and 0500 is 5 p.m. Local time is then converted to Zulu or GMT with the appropriate conversion for the time zone the flight mission is taking place.

Answer – b. Aviation uses the 24-hour clock (as explained in the answer) that then is converted to GMT or Zulu time. Reference: PHAK, AIM, AWH

177. Refer to Figure 27. An Automated Flight Service Station (AFSS) briefer ask for the time of your flight mission. You plan on beginning the mission at 0700 Eastern daylight time (local). What is the time when converted to Zulu?

a) 1100Z

b) 1200Z

c) 0200Z

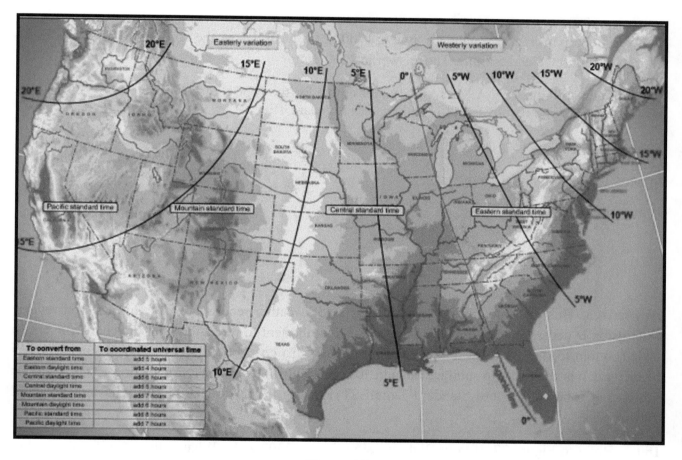

Figure 27

Answer – a. 0700 or 7 a.m. local, Eastern daylight time requires a conversion to Zulu time (a.k.a. Greenwich Mean Time-GMT). In Figure 28, the bottom left side of the chart shows a conversion by adding 4 hours. 0700 plus 4 hours equals 1100 Zulu time.

178. Refer to Figure 27. An Automated Flight Service Station (AFSS) briefer, asked for the time of your flight mission. You plan on beginning the mission at 5 p.m. Eastern Standard Time. What is the time when converted to Zulu?

a) 1200 Zulu

b) 2200 Zulu

c) 0400 Zulu

Answer – b. 5 p.m. Eastern Standard time is 1700 hours local. Using the conversion chart on the left bottom of Figure 28 add 5 hours. The answer is 2200 Zulu.

179. Refer to Figure 27. A METAR weather report for the KGSO airport indicates SPECI 1205Z. What is the time if converted to local Eastern Standard Time?

a) 0705 Eastern Standard Time.

b) 0805 Eastern Standard Time.

c) 0905 Eastern Standard Time.

Answer – a. This question can be a little tricky since weather reports are in Zulu time and reverse conversion is necessary. When using Figure 28 it is important to remember that the chart conversion table (bottom left) is for local time conversion to Zulu. To go from Zulu, what the weather report is being used, it is necessary to subtract, not add, to covert Zulu to local. Therefore, taking 1205 Z requires subtracting 5 hours (Eastern Standard Time-EST) making the answer 0705 hours Eastern Standard Time (EST).

180. Refer to Figure 27. If your flight mission time is 2300 Zulu. What is the time if converted to Mountain Standard Time (MST)?

a) 1600 MST or 5 p.m.

b) 1600 MST or 4 p.m.

c) 0500 MST or 7 a.m.

Answer – b. When referring to the Figure 27 graphic on the bottom left side, it is important to remember that this is for Local to Zulu time conversions. Therefore, it is necessary to subtract when going from Zulu to Local. Subtract 7 hours from 2300 Z, the answer is 1600 MST or 4 p.m. Reference: PHAK, AWH, AIM

181. To get a complete weather briefing for the planned mission flight the Remote Pilot (RP) should request

a) a general briefing.

b) an abbreviated briefing.

c) a standard briefing.

Answer – c. A Standard Weather Briefing is a comprehensive briefing that will give a pilot an overview of regional conditions down to the current weather and forecast condition of the nearest airport with weather reporting capabilities. Additionally, the briefer will relay NOTAMS and TFR's. A briefing can be done via the Automated Flight Service Station at, 1-800-WX-BRIEF, or online at 1800WXBrief.com. This is considered a "legal briefing" (complies with FAA pre-flight regulations) by the FAA. Reference: AWH

182. Which type of weather briefing should a Remote Pilot (RP) request when the mission flight will be taking place within the hour if no preliminary weather information has been received?

a) Outlook briefing.

b) Abbreviated briefing.

c) Standard briefing.

Answer – c. A Standard Weather Briefing is a comprehensive briefing that will give a pilot an overview of regional conditions down to the current weather and forecast condition of the nearest airport with weather reporting capabilities. Additionally, the briefer will relay NOTAMS and TFR's. A briefing can be done via the Automated Flight Service Station at, 1-800-WX-BRIEF, or online at 1800WXBrief.com. This is considered a "legal briefing" (complies with FAA pre-flight regulations) by the FAA. Reference: AWH

183. What type of weather briefing should a Remote Pilot (RP) request when the flight mission time is greater than six hours away?

a) An outlook briefing.

b) A supplemental briefing.

c) An abbreviated briefing.

Answer – a. According to the FAA, "You should request an Outlook Briefing whenever your proposed time of departure is six or more hours from the time of the briefing. The briefer will provide available forecast data applicable to the proposed flight. This type of briefing is provided for planning purposes only. You should obtain a Standard or Abbreviated Briefing prior to departure…" Reference: AWH, http://www.faa.gov/about/office_org/headquarters_offices/ato/service_units/systemops/fs/alaskan/alaska/fai/pfpwb/.

184. What should a Remote Pilot (RP) state initially when telephoning a weather briefing facility, Automated Flight Service Station (AFSS), for preflight weather information?

a) Tell the FAA FSDO they report.

b) State their total flight time.

c) Identify themselves as an RP.

Answer – c. The standard protocol when contacting an AFSS (1-800 WX BRIEF or 1800WXBRIEF.com) is to identify oneself as an RP. This is typically followed by the aircraft's registration number issued by the FAA. Reference: AWH

185. When telephoning a weather briefing facility for preflight weather information the Remote Pilot (RP) should state

a) the aircraft identification or their name.

b) the maximum average speed at which the mission will be flown.

c) type of fuel used, battery or standard fuel, and total available flight duration time in tenths of an hour.

Answer – a. The standard protocol when contacting an AFSS (1-800 WX BRIEF or 1800WXBRIEF.com) is to identify oneself as an RP. This is typically followed by the aircraft's registration number issued by the FAA. Reference: AWH

186. To update a previous weather briefing from the Automated Flight Service Station (AFSS) a Remote Pilot (RP) should request

a) an abbreviated briefing.

b) a standard briefing.

c) an outlook briefing.

Answer – a. Weather briefing is obtained by contacting an AFSS (1-800 WX BRIEF or 1800WXBRIEF.com). Outlook briefings are done well before mission time (greater than 6 hours). Standard weather briefings are a full comprehensive briefing done closer to mission time. Abbreviated Briefings are accomplished just prior to the mission time (an hour prior for example) to recheck items such as TFR's, changing weather, etc. An AFSS briefing is considered a "legal briefing" (complies with FAA pre-flight regulations) by the FAA. Reference: AWH

187. When requesting weather information for the following morning a Remote Pilot (RP) should request

a) an outlook briefing.

b) a standard briefing.

c) an abbreviated briefing.

Answer – a. Outlook briefings are done well before mission time (greater than 6 hours). Standard weather briefings are a full comprehensive briefing done closer to mission time. Abbreviated Briefings are accomplished just prior to the mission time (an hour prior for example) to recheck items such as TFR's, changing weather, etc. AFSS briefings are considered a "legal briefing" (complies with FAA pre-flight regulations) by the FAA. Reference: AWH

188. You plan to phone a weather briefing facility (Automated Flight Service Station) for preflight weather information. You should

a) provide the phone number for emergency contact information.

b) identify yourself as a Remote Pilot (RP).

c) begin with your flight mission area.

Answer – b. The standard protocol when contacting an AFSS (1-800 WX BRIEF or 1800WXBRIEF.com) is to identify oneself as an RP. This is typically followed by the aircraft's registration number issued by the FAA. Reference: AWH

189. In aviation weather, a cloud ceiling is defined as the height above the Earth's surface of the

a) lowest reported obscuration and the highest layer of clouds reported as overcast.

b) lowest broken or overcast layer of vertical visibility into an obscuration.

c) lowest layer of clouds reported as scattered, broken, or thin.

Answer – b. The covering of the sky by clouds is described from the fewest (or no) clouds to total coverage (see graphic below). Respectively, this coverage is expressed as; Sky Clear (SKC), Few (FEW), Scattered (SCT), Broken (BKN) and Overcast (OVC). However, FEW and (SCT) clouds are not considered a cloud ceiling in aviation weather.

The answer uses unclear wording. "Vertical visibility," means looking straight up into the sky. "Into and obstruction," is looking at a cloud. The cloud layer closest to the Earth must be deemed Broken or Overcast in-order to meet the definition of a cloud ceiling. Reference: AWH

Sky Cover	Contraction
Less than ⅛ (Clear)	SKC, CLR, FEW
⅛–2/8 (Few)	FEW
⅜–4/8 (Scattered)	SCT
⅝–⅞ (Broken)	BKN
8/8 or (Overcast)	OVC

190. If the temperature-dew point spread is small and decreasing, and the temperature is 62 degrees F, what type weather is most likely to develop?

a) Freezing precipitation.

b) Thunderstorms.

c) Fog or low clouds.

Answer – c. Freezing precipitation is out of the question with a 60° F temperature as in answer a. Thunderstorms are a close second, but there is no mention of the stability for air. Fog or low clouds is the best answer because it is considered visible moisture, creating fog (clouds at ground level) or low clouds. Reference: AWH

191. What causes variations in barometric pressure between weather reporting points?

a) Unequal heating of the Earth's surface.

b) Variation in terrain elevation.

c) Coriolis Force.

Answer – a. The weather is caused by the unequal heating of the Earth's surface which results in varying barometric pressure readings across reporting stations. Higher temperatures mean lower air density and corresponding lower pressure. Reference: AWH

192. What causes the wind at higher altitudes (i.e. 3,000 AGL) to be different than the wind at the surface?

a) stronger pressure gradient at higher altitudes.

b) friction between the wind and the surface.

c) stronger Coriolis Force at the surface.

Answer – b. The change in wind direction from the surface to altitude (above the surface) is due to Surface Friction. As wind interacts with the surface of the Earth, it changes course. This can be verified by comparing the surface wind to Winds Aloft reports. Reference: AWH

193. The development of thermals depends on

a) a counterclockwise circulation of air.

b) temperature inversions.

c) solar heating.

Answer – c. Flying when thermal activity is the greatest, typical afternoon, can be the most challenging to fly. Many people agree that an early morning or late afternoon is better for reduced turbulence associated with the thermal activity. This hour close to sunrise and sunset is called "The Golden Hour for its dynamic hues. Reference: AWH

194. You have received an outlook briefing from flight service through 1800wxbrief.com. The briefing indicates you can expect a low-level temperature inversion with high relative humidity. What weather conditions would you expect?

a) Smooth air, poor visibility, fog, haze, or low clouds.

b) Light wind shear, poor visibility, haze, and light rain.

c) Turbulent air, poor visibility, fog, low stratus type clouds, and showery precipitation.

Answer – a. A low-level temperature inversion is a low level of air that has temperatures rising, instead of declining, as altitude goes up. One way to remember this is to think about characteristics of normal temperature as it decreases when driving up a mountain to greater altitudes. Combined with high humidity is a formula for smooth air, poor visibility, fog, haze, or low clouds. References: FAA Advisory Circular AC 00-6A - Aviation Weather For Pilots and Flight Operations Personnel, Q35UAS SG

195. What are the characteristics of a moist, unstable air mass?

a) Turbulence and showery precipitation.

b) Poor visibility and smooth air.

c) Haze and smoke.

Answer – a. An unstable air mass will typically have cumulous clouds and turbulence with the associated instability. As the cumulous clouds gain strength through upward and downward movement of unstable air, it produces showery precipitation. References: Advisory Circular AC 00-6A - Aviation Weather For Pilots and Flight Operations Personnel, Q37UAS SG

196. What are the characteristics of stable air?

a) Good visibility and steady precipitation.

b) Poor visibility and steady precipitation.

c) Poor visibility and intermittent precipitation.

Answer – b. Stable air typically forms clouds that don't have a lot of vertical development (like cumulous clouds). While this makes for good stable air for flying visibility can be poor with steady precipitation due to the uniform cloud structure. Stratus clouds are a good example of stable air clouds. References: Advisory Circular AC 00-6A - Aviation Weather For Pilots and Flight Operations Personnel, Q38UAS SG.

197. Refer to Figure 12: The wind direction and velocity at KJFK is from

a) 180° true at 4 knots.

b) 180° magnetic at 4 knots.

c) 040° true at 18 knots.

METAR KINK 121845Z 11012G18KT 15SM SKC 25/17 A3000

METAR KBOI 121854Z 13004KT 30SM SCT150 17/6 A3015

METAR KLAX 121852Z 25004KT 6SM BR SCT007 SCT250 18/15 A2991

SPECI KMDW 121856Z 32005KT 1 1/2SM RA OVC007 17/16 A2980 RMK RAB35

SPECI KJFK 121853Z 18004KT 1/2SM FG R04/2200 OVC005 20/18 A3006

FIGURE 12.—Aviation Routine Weather Reports (METAR).

Answer – a. The wind is reported in knots and via true north. In this question, the reference is the following: SPECI KJFK 121853Z 18004KT 1/2SM FG R04/2200 OVC005 20/18 A3006. This is a special report (SPECI), meaning a significant change in the weather has been reported. 121853Z: The date is the 12th of the month, time 1853 Zulu. 18004KT: Winds are **from** the south, 180 degrees (true north) at 4 knots. 1/2SM FG: 1/2 Statute Mile visibility with Fog. R04/2200. – Runway 4 Runway Visual Range is 2,200' (visibility down the runway). Clouds are Overcast at 500' Above Ground Level (AGL). 20/18: Temperature 20 degrees C, Dew Point 18. A3006: Altimeter barometric setting 30.06. References: AviationWeather.gov product information; https://aviationweather.gov/products/nws/info, AC 00-6A - Aviation Weather For Pilots and Flight Operations Personnel, FAA-CT-8080-2G, Q39UAS SG

198.   Refer to Figure 12: What are the current conditions for Chicago Midway Airport (KMDW)?

a) Sky 700 feet overcast, visibility 1-1/2SM, rain.

b) Sky 7000 feet overcast, visibility 1-1/2SM, heavy rain.

c) Sky 700 feet overcast, visibility 11, occasionally 2SM, with rain.

Answer –a. The Midway weather is listed as a Special (SPECI) due to a significant change in the weather. The report is: SPECI KMDW 121856Z 32005KT 1 1/2SM RA OVC007 17/16 A2980 RMK RAB35. 121856Z: The date is the 12th of the month, time 1856 Zulu. 32005KT: Winds are 320 degrees (true north) at 5 Knots. 1 1/2SM RA: Visibility is 1 1/2 Statute Miles with RAin. OVC007: Clouds are Overcast at 700' Above Ground Level (AGL). References: AC 00-6A - Aviation Weather For Pilots and Flight Operations Personnel, FAA-CT-8080-2G, Q40UAS SG.

199.   Refer to Figure 12: The remarks section for KMDW has RAB35 listed. This entry means

a) blowing mist has reduced the visibility to 1-1/2 SM.

b) rain began at 1835Z.

c) the barometer has risen 0.35 Hg.

Answer – b. RAB35, at the very end of the SPECI (METAR) means that Rain Began (RAB) at 35 minutes past the hour. Reference: AWH

200. Figure 12: What are the wind conditions at Wink, Texas (KINK)?

a) Calm.

b) 110 degrees at 12 knots, gusts 18 knots.

c) 111 degrees at 2 knots, gusts 18 knots.

Answer – b. The report for wind direction and wind speed is in the grouping 11012G18KT. Translated means winds are heading from 110 degrees true at 12 knots, gusts to 18 knots. Reference: AWH

201. What are Pilot Reports (a.k.a. PIREP's) and what are they used for?

a) PIREP's are reports from pilots stating weather conditions for their specific location and distributed for other pilots to gain information. In a weather report, a PIREP begins with the letters UA.

b) Pilot Reports are reports of uncharacteristic smoke patterns that may obstruct visibility at certain times of the year.

c) Pilot Reports are surveys or reports made by pilots to Automated Flight Service Stations (AFSS) management to improve the algorithms of weather data by comparing forecast with actual conditions for improved reporting systems.

Answer – a. If you happen to run across PIREP's (UA's) while checking the weather online know that they are Pilot Reports about in-flight weather conditions. Reference: AWH

202. To best determine general forecast weather conditions covering a multi-State region the Remote Pilot (RP) should refer to

a) aviation area forecasts.

b) weather depiction charts.

c) satellite maps.

Answer – a. Area forecast provides an overview of weather over a multi-State region. See the areas covered and more at http://www.aviationweather.gov/products/fa/. Reference: AWH

203. The section of the Area Forecast entitled "VFR CLDS/WX" contains a general description of

a) cloudiness and weather significant to flight operations broken down by states or other geographical areas.

b) forecast sky cover, cloud tops, visibility, and obstructions to vision along specific manned aircraft routes.

c) clouds and weather which cover an area greater than 3,000 square miles and is significant to Visual Flight Rules VFR flight operations.

Answer – c. The Area Forecast is a multi-State overview of forecasted weather. Deconstructing the question, VFR, or Visual Flight Rules, is good weather flying, CLDS stands for "clouds" and WX is short for "weather" and you have a multi-State forecast for clouds for VFR pilots. The only thing you have to remember is that this particular forecast is for "an area greater than 3,000 square miles." Reference: AWH

204. From which primary source should information be obtained regarding the expected weather at the end of a mission day if your closest airport with weather reporting capabilities does not have a Terminal Forecast?

a) Low-Level Prognostic Chart.

b) Weather Depiction Chart.

c) Area Forecast.

Answer – c. The Area Forecast (AF) is a multi-State overview of forecasted weather. In the absence of a Terminal Forecast the AF is a good option. Low Level Prognostic Chart is a good answer as well, but the FAA is leaning to the AF over the "Prog Chart." Reference: AWH

205.   Figure 15: In the TAF for KMEM what does "SHRA" stand for?

a) Rain showers.

b) A shift in wind direction is expected.

c) A significant change in precipitation is possible.

```
TAF

KMEM 121720Z 1218/1324 20012KT 5SM HZ BKN030 PROB40 2022 1SM TSRA OVC008CB
     FM2200 33015G20KT P6SM BKN015 OVC025 PROB40 2202 3SM SHRA
     FM0200 35012KT OVC008 PROB40 0205 2SM-RASN BECMG 0608 02008KT BKN012
     BECMG 1310/1312 00000KT 3SM BR SKC TEMPO 1212/1214 1/2SM FG
     FM131600 VRB06KT P6SM SKC=

KOKC  051130Z 0512/0618 14008KT 5SM BR BKN030 TEMPO 0513/0516 1 1/2SM BR
      FM051600 18010KT P6SM SKC BECMG 0522/0524 20013G20KT 4SM SHRA OVC020
      PROB40 0600/0606 2SM TSRA OVC008CB BECMG 0606/0608 21015KT P6SM SCT040=
```

FIGURE 15.—Terminal Aerodrome Forecasts (TAF).

Answer – a. You don't necessarily have to look at Figure 15 to decode SHRA but it is good to practice translating many Terminal Area Forecast (TAF). SH-RA seems like it is backward because SH means "showers" and RA stands for "rain." Reverse and put them together and you get Rain Showers (SHRA). Wish I could find the guy that came up with that one. Reference: AWH

206.   During the time period from 0600Z to 0800Z what visibility is forecast for KOKC, Figure 15?

a) Greater than 6 Statute Miles (SM).

b) Possibly 6 Statute Miles (SM).

c) Not forecasted.

Answer – a. When visibility is not given in a Terminal Area Forecast, it is assumed that visibility is greater than 6 SM. Reference: AWH

207. Figure 15: In the TAF from KOKC the clear sky becomes

a) overcast at 2,000 feet during the forecast period between 2200Z and 2400Z.

b) overcast at 200 feet with a 40 percent probability of becoming overcast at 600 feet during the forecast period between 2200Z and 2400Z.

c) overcast at 200 feet with the probability of becoming overcast at 400 feet during the forecast period between 2200Z and 2400Z.

Answer – a. In this question look for the BECMG (Becoming) mid page, second line. On the 5th of the month from 22Z to 24Z the forecast is for 4 SM visibility, Rain Showers (SHRA), Overcast (OVC) cloud ceiling at 2000' AGL. Reference: AWH

208. What is the valid period for the TAF for KMEM when you see at the beginning of the report, "KMEM 121720Z 121818...?"

a) 1200Z to 1200Z.

b) 1200Z to 1800Z.

c) 1800Z to 1800Z.

Answer – c. The Terminal Area Forecast (TAF) covers a 24-hour time frame. Looking at the first section of numbers from the KMEM TAF you can see it was made on the 12th of the month at 1720Z (121720Z). 1218 decoded is again the 12th of the month and the forecast begins at 18Z and is good until 18Z, 24-hours later. Reference: AWH

209. Between 1000Z and 1200Z the visibility at KMEM is forecast to be (Figure 15)?

a) 1/2 Statute Mile (SM).

b) 3 Statute Miles (SM).

c) 6 Statute Miles (SM).

Answer – b. The Terminal Area Forecast begins at 18Z on the 12th of the month (KMEM 121720Z **1218**). Skipping down to the fourth line of text you see 1310/1312 which means, on the 13th of the month from 10Z to 12Z followed by calm winds (0000KT). Next, the visibility is forecast to be 3SM in Mist (BR), 3SM is the answer.

210. Figure 15: What is the forecast wind for KMEM from 1600Z until the end of the forecast?

a) Zero wind.

b) Variable at 6 knots.

c) Variable at 4 knots.

Answer – b. If you jump down to the last line of the KMEM Terminal Area Forecast (TAF), notice FM121600 VRB06KT which means From (FM) the 13th of the month at 16Z the winds are forecast to be Variable (in wind direction) at 6 Knots (KTS) with visibility greater than (P) 6 SM with Clear Skies (SKC). Reference: AWH

211. Figure 15: In the TAF from KOKC, the "FM (FROM) Group" is forecast for the hours from 1600Z to 2200Z with the wind from

a) 160 degrees at 10 knots.

b) 180 degrees at 10 knots.

c) 180 degrees at 10 knots becoming 200 degrees at 13 knots.

Answer – b. On the second line you see FM051600 18010KT P6SM SKC BECMG 0522…" which means From (FM) the 5th of the month at 16Z that the winds are forecast to be from 180° at 10 Knots (KTS) with visibility greater than (P) 6 SM with the Sky Clear (SKC), then Becoming (BECMG) on the 5th of the month from 22Z…. Reference: AWH

212. Figure 15: The only cloud type forecast in TAF reports is

a) Nimbostratus.

b) Cumulonimbus.

c) Scattered Cumulus.

Answer – b. The only type of clouds forecast in a TAF, are Cumulonimbus. Reference: AWH, QR

213.   Figure 18: What is the value of the Weather Depiction Chart to the Remote Pilot (RP)?

a) For determining general weather conditions over a large area.

b) For a forecast of cloud coverage, visibilities, and frontal activity.

c) For determining frontal trends and air mass characteristics.

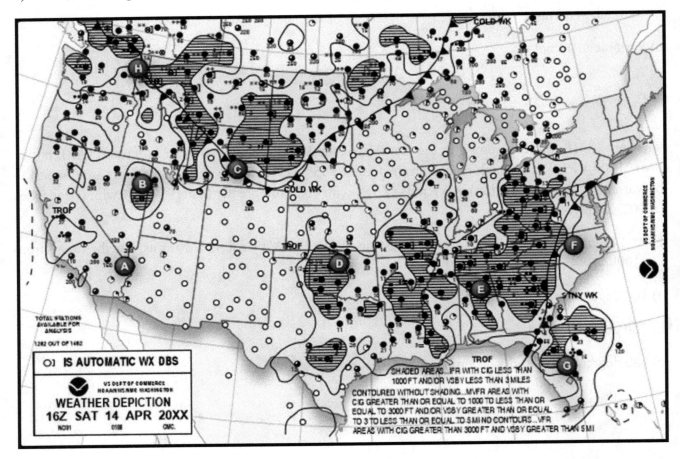

Figure 18

Answer – a. Weather Depiction Charts do not show fronts, METARs, TAFs, etc., but are a good general planning guide for overall weather. The note in the Gulf of Mexico best states the specific attributes of the Weather Depiction Chart. This chart could be used for a quick check of the weather over a State or for multi UA operations over a region. Reference: AWH

214. Radar weather reports are of special interest to sUAS flight crews because they indicate

a) large areas of low ceilings and fog.

b) location of precipitation along with type, intensity, and cell movement of precipitation.

c) location of precipitation along with type, intensity, and trend.

Answer – b. Aviation radar weather reports are just like the images seen on the nightly new weather segment with areas of green, yellow, red and purple to indicate intensity. There is the ability to tell the type of precipitation in addition to cell movement.

Tip: Ensure the weather service you use has as close to up to the minute depictions. Some radar reports can be up to 15 minutes old before it reaches your display device. Reference: AWH

215. Figure 19: What weather is forecast for the Florida area just ahead of the stationary front during the first 12 hours?

a) Ceiling 1,000 to 3,000 feet and/or visibility 3 to 5 miles with intermittent precipitation.

b) Ceiling 1,000 to 3,000 feet and/or visibility 3 to 5 miles with continuous precipitation.

c) Ceiling less than 1,000 feet and/or visibility less than 3 miles with continuous precipitation.

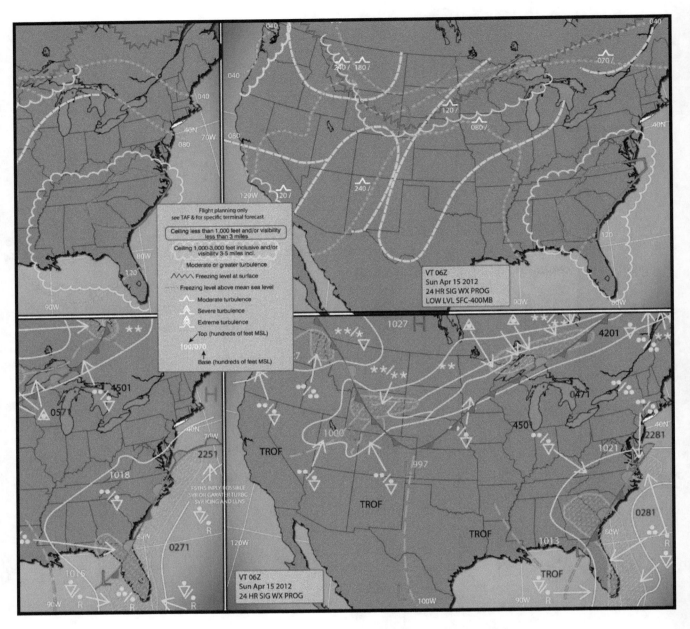

Figure 19 Low Level Significant Weather Prognostic Chart.
Note: Partial chart, left side is 12 hour, right side 24 hour.

Answer – b. The left-hand side of the Weather Prog Chart is the 12-hour view; the right-hand side is the 24-hour forecast. Looking at the top left image you see light blue (cloud outline like) lines that in case the whole Southeast US. Using the Figure 18 inset to decode the blue lines it states that ceilings are forecast to be 1,000' – 3,000' and visibility from 3 – 5 miles. Reference: AWH

216. Figure 19: Interpret the weather symbol depicted in Utah on the 12-hour (left-hand side) Significant Weather Prognostic Chart.

a) Moderate turbulence, surface to 18,000 feet.

b) Thunderstorm tops at 18,000 feet.

c) The base of clear air turbulence, 18,000 feet.

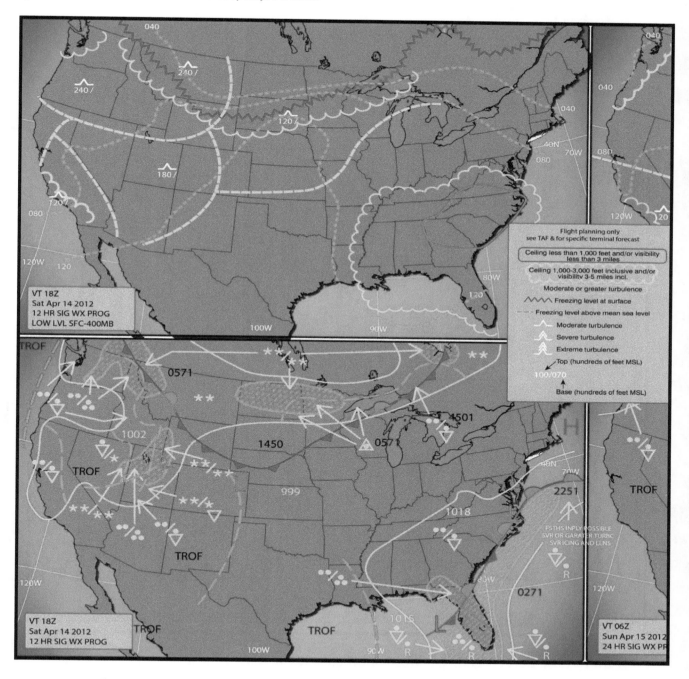

Figure 19 Low Level Significant Weather Prognostic Chart (12 hour, left hand side)

Answer – a. Looking at the Utah region, you can see the dashed yellow lines (upper and lower charts) which mean turbulence (see inset). Also over the State is the Moderate Turbulence symbol (see inset) with the number 180 directly below, meaning Moderate Turbulence 18,000' MSL and below. Reference: AWH

217.   Figure 19: The enclosed shaded area associated with the low-pressure system over northern Utah is forecast to have

a) continuous snow.

b) intermittent snow.

c) continuous thunderstorms.

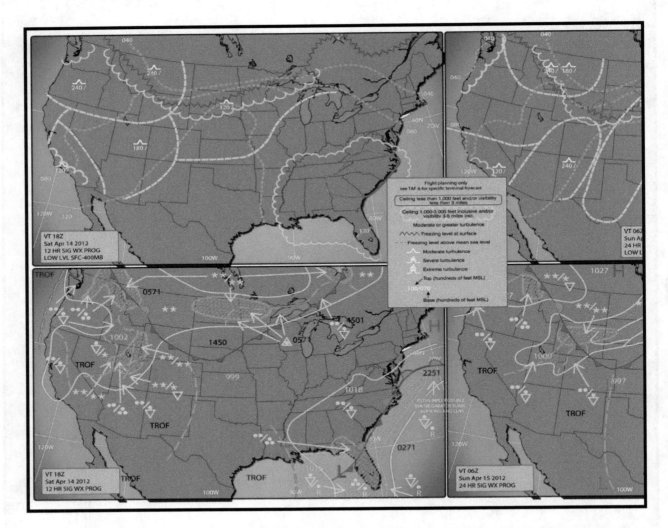

Figure 19 Low-Level Significant Weather Prognostic Chart. Graphic modified to show clarity.

Answer – a. This question can be tricky because of the mass of symbols over the Utah area. The symbol being referenced in the question are the two stars with a slash followed by two stars (**/**) which, decoded, means continuous snow. Reference: AWH

218. Figure 19: At what altitude is the freezing level over the middle of Florida on the 12-hour Significant Weather Prognostic Chart?

a) 4,000 feet.

b) 8,000 feet.

c) 12,000 feet.

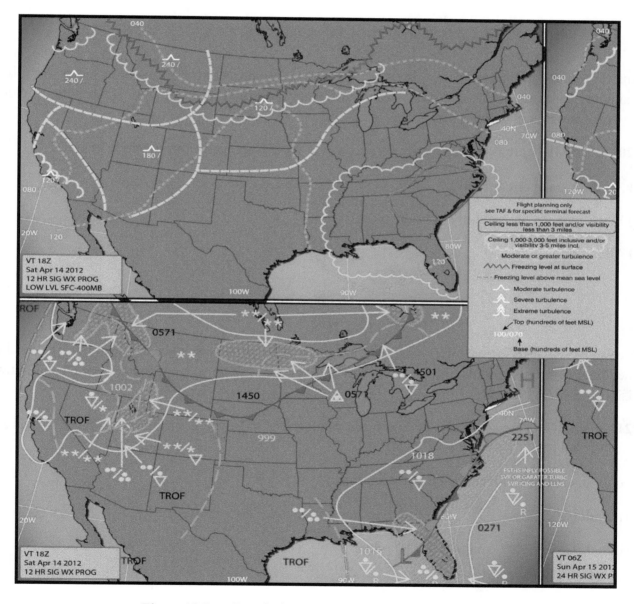

Figure 19 Low-Level Significant Weather Prognostic Chart

Answer – c. The inset shows that a green dashed line indicates the freezing level. Looking at the top chart, there is a green line that cross through the middle of Florida with a 120 under the dashed line meaning the freezing level is 12,000' MSL. Reference: AWH

219.   How are Significant Weather Prognostic Charts best used by a sUAS flight crew?

a) For submission to the FAA for special flight permits.

b) For determining areas to avoid (freezing levels and turbulence).

c) For analyzing current frontal activity and cloud coverage.

Answer – b. Significant Weather Prognostic Charts ("Prog Charts) are a good resource for flight planning that contains freezing levels and turbulence information in addition to beneficial cloud coverage and precipitation types. Reference: AWH

220.   SIGMETs are issued as a warning of weather conditions hazardous to which aircraft?

a) Small manned aircraft only.

b) Large manned aircraft only.

c) All aircraft.

Answer – c. SIGMETs give pilots information about **severe**; turbulence and icing in addition to weather phenomenon such as; sandstorms, dust-storms and volcanic ash episodes that are reported as limiting visibility to 3 SM or less.

221.   AIRMETs are advisories of significant weather, but of lower intensities than SIGMETs and are intended for dissemination to

a) only Instrument (IFR) manned aircraft pilots.

b) all pilots.

c) only Visual Flight Condition (VFR) pilots.

Answer – b. Memory aid: SIGMETs are used for SIGnificant weather issues, AIRMETs are for less severe weather for all general AIRMEn. Never-the-less, all pilots can benefit from AIRMETs. AIRMETs cover less severe weather like; moderate turbulence and icing sustained surface winds of 30 knots or more, or widespread restricted visibility. Reference: AWH

222.   Which weather advisory would contain information on severe icing not associated with thunderstorms?

a) Convective SIGMET.

b) SIGMET.

c) AIRMET.

Answer – b. SIGMETs give pilots information about **severe**; turbulence and icing in addition to weather phenomenon such as; sandstorms, dust storms and volcanic ash episodes that are reported as limiting visibility to 3 SM or less. Looking at the answers, a Convective SIGMET would not be the best because it is describing convective (thunderstorm) activity. AIRMETs do not cover severe weather, and therefore, a standard SIGMET is the best answer. Reference: AWH

223.   What information is contained in a CONVECTIVE SIGMET?

a) Tornadoes, embedded thunderstorms, and hail 3/4 inch or greater in diameter.

b) Severe icing, severe turbulence, or widespread dust storms lowering visibility to less than 3 miles.

c) Surface winds greater than 40 knots or thunderstorms equal to or greater than video integrator processor (VIP) level 4.

Answer – a. Convective means thunderstorms and SIGMETs convey areas of Severe weather, such as thunderstorms and their associated spin-offs such as hail and tornadoes. Answer b doesn't include thunderstorms. While answer c mentions thunderstorms, the VIP level 4 isn't relevant to the answer. Reference: AWH

224. What is indicated when a current CONVECTIVE SIGMET, forecast thunderstorms?

a) Moderate thunderstorms covering 30 percent of the area.

b) Moderate or severe turbulence.

c) Thunderstorms obscured by massive cloud layers.

Answer – c. Convective (thunderstorm) SIGMETs are reported if severe or greater, therefore answers a and b are incorrect. The best answer is c since thunderstorms can be embedded and "obscured by massive cloud layers." Reference: AWH

225. The Automatic Terminal Information Service (ATIS) is the continuous broadcast of recorded information concerning

a) pilots of radar-identified aircraft whose aircraft is in dangerous proximity to terrain or an obstruction.

b) nonessential information to reduce frequency congestion.

c) non-control information in selected high-activity terminal areas.

Answer – c. The question is very wordy and confusing in our opinion. This being the case, the best way to remember this is by breaking down the terminology. ATIS is a pre-recorded weather reporting for the immediate airport area stating; winds, cloud ceilings, visibility, barometric pressure, runway/taxiway closures, etc. Breaking down the answer, "non-control information" means information that is not normally part of air traffic control duties per say (ATIS is pre-recorded by an automated voice, and a system, affectionately called "Perfect Paul." Followed by the statement, "in selected high-activity terminal areas," is a way of saying ATIS exist at busy airports. Reference: PHAK, AIM

226. The absence of the sky condition and visibility on an ATIS broadcast indicates that

a) weather conditions are at or above VFR minimums.

b) the sky condition is clear, and visibility is unrestricted.

c) the ceiling is at least 5,000 feet, and visibility is 5 miles or more.

Answer – c. The ATIS reports offer a maximum of reporting capabilities of cloud ceilings up to 5,000' AGL and visibility to 5 miles, anything greater is not required to be reported. Since the function of ATIS is to give weather information in the immediate airport area, think of ATIS as serving the exact function of its definition. References: PHAK, AIM

Flight Performance: accessories and loads

227. To ensure that the unmanned aircraft's center of gravity (CG) limits is not exceeded, follow the aircraft loading instructions specified in the

a) Pilots Operating Handbook (POH) or Manufacturer's owner's manual or in the absence of manufacturer data use industry best practices or other creditable data.

b) Aeronautical Information Manual (AIM).

c) FAA Aircraft Weight and Balance Handbook.

Answer – a. Details of CG limitations are specific to the make and model of UAS as detailed by the manufacturer. Manufacturer data would contain manufacturer information about weight and balance (CG) limitations. b-The AIM does not contain CG/Weight & Balance information. c - The FAA Aircraft Weight and Balance Handbook is an overview and not specific to a make and model of UAS. Reference: Q8UAS SG.

It Happens: Emergency Procedures

Reporting Accidents/Incidences

228. To avoid a possible collision with a manned airplane you estimate that your small UA climbed to an altitude greater than 600 feet AGL. To whom you must report the deviation?

a) Air Traffic Control.

b) The National Transportation Safety Board.

c) Upon request of the Federal Aviation Administration.

Answer – c. This incident does not meet the guidelines in an accident in CFR 107.9 since there was no damage to property or injury. A climb above 400' AGL may or may not have happened in Air Traffic

Controlled airspace, so it might not be appropriate to report it to Air Traffic Control. Qualifying accident/incident in manned aircraft is reported to the National Transportation Safety Board (NTSB). However, the protocol for sUAS is to report to the Federal Aviation Administration (FAA). Since this does not meet the damage/injury to persons or property the FAA may not request a formal report.

Tip: The NTSB operates as an unbiased third party between pilots and the FAA for the voluntary reporting of infractions. If the RPIC knows a regulation was broken, a report, called a NASA Form, can be filled out describing the details surrounding the incident. This is done by many pilots in exchange for leniency **if** the FAA were to pursue the issue with the pilot. The NASA Form should be completed shortly after the flight in question prior to receiving FAA correspondence. References: Q14UAS SG

229. An accident involving a sUAS crew involving serious injury must be reported to the FAA within

a) 24 hours of the accident.

b) 48 hours of the accident.

c) 10 calendar days of the accident.

Answer – c. Accidents involving UA are unique in the sense of who handles the reporting and under what timeline. Manned aircraft report to the NTSB under a different timeline. The FAA now has the responsibility for UA accidents and requires reporting within 10 calendar days to the FAA FSDO or FAA Regional Reporting Office. Reference: CFR 107.9

230. During a sUAS flight mission, the aircraft experiences a very hard landing due to a rouge wind gust and subsequent loss of control. Fortunately, no persons or property were damaged, but the aircraft was a total loss (a $1,500 replacement cost). As a result, the Remote Pilot (RP) is required to report the accident to the

a) National Transportation Safety Board (NTSB) within 48 hours of the accident.

b) Federal Aviation Administration (FAA) within 10 days of the accident.

c) No reporting is required since there was no injury to persons or damage to property meeting the respective threshold amounts. Damage or total loss of the sUAS is not calculated in the FAA definition of an accident.

Answer – c. To be deemed a reportable accident, according to the FAA injury to person(s) is required to be a Level 3 or higher injury. Property damage amounts, not counting the sUAS are required to be reported if to be "greater than $500 to repair or replace the property (whichever is less). Reference: CFR 107.9, AC 107-2 (4.5)

231.   During a sUAS flight mission on a film set, with a waiver to fly over people, a lone actor is running across a grassy field as the aircraft follows in a chase sequence. The actor trips over un-level ground and momentarily loses consciousness. He is awake and alert before on-site medical personnel arrives on the scene but is taken to the closest hospital for evaluation. Does this situation warrant notification of the FAA and if so within what time frame?

a) No. Since the accident didn't have anything to do directly with the sUAS there is no reason to report the situation to the FAA.

b) Yes. The regulations stipulate that if any person has a loss of consciousness during a sUAS flight mission, then a report has to be made to the NTSB within 10 days of the accident.

c) Yes. The regulations stipulate that if any person has a loss of consciousness during a sUAS flight mission, then a report has to be made to the FAA within 10 calendar days of the accident.

Answer – c. This scenario is reportable as an accident (within 10 calendar days) because it involved a person, or as FAA guidance states "any person," that has a serious injury (Level 3 or greater) **or** any loss of consciousness. "Any person" doesn't just mean the sUAS flight crew and is more extensive than those involved with the flight mission. Reference: CFR 107.9, AC 107-2 (4.5)

232.   An accident happens on the location of a sUAS flight mission in which the flight mission aircraft crashes into a customer's new fleet sedan automobile resulting in $327 in damage to a headlight assembly. What FAA reporting is required by the Remote Pilot (RP)?

a) No action is required as far as the FAA is concerned, the cost of repair of the automobile does not exceed $500.

b) A report must be made to the FAA because the material and labor cost are more than $300.

c) No action is required as far as the FAA is concerned because the cost of repair does not exceed $1,000.

Answer – a. According to the FAA guidance, "if the cost is greater than $500 to repair or replace the property (whichever is lower)." In this case, the repair cost is under the $500 threshold. Reference: CFR 107.109

233. During a film shoot of closed set surfing demonstration, a sUAS crew accidentally loses control of the sUAS sending it crashing into a high-performance surfboard placed upright on a palm tree as a backdrop for surfer interviews. Fortunately, nobody was hurt during the accident, but a store owner said the board was a total loss due to damage was done by the aircraft. The board, just prior to the accident, according to the store owner the board was worth $1,200. Does this accident have to be reported to the FAA?

a) No. Since no serious injury or loss of life resulted in the accident further action is not necessary. However, a notation must be made in the aircraft maintenance logs reporting the accident.

b) Yes. According to FAA regulations, the fair market value of the property exceeds $500 as a total loss and therefore must be reported to the FAA.

c) Yes. According to FAA regulations, the fair market value of the property exceeds $1,000 as a total loss and therefore must be reported to the FAA.

Answer – b. The cost of replacement or repair, in this case, replacement, exceeds the $500 threshold, it must be reported to the FAA FSDO or Regional Operations Center (ROC). Reference: CFR 107.9

234. How many days does the Remote Pilot (RP) of a sUAS have to report an accident to the FAA?

a) 14 business days.

b) 10 business days.

c) 10 calendar days.

Answer – c. You might think at first glance reporting would be on a business day cycle. However, the actual requirement set by the FAA is 10 calendar days. Reference: AC 107-2 (4.5)

235. To be classified as an FAA defined sUAS accident, an injury would have to be classified as

a) a Level 3 (serious) or higher based on the Abbreviated Injury Scale (AIS) or involve any loss of consciousness.

b) an accident by the Remote Pilot (RP) for reporting purposes.

c) an accident if First Responders are contacted and respond.

Answer – a. A Level 3 or greater injury is classified as Serious (Level 3) to Death (Level 6). However, it is also important to note that any loss of consciousness is reportable as well. Reference: AC 107-2 (4.5)

236. Your sUAS damages property whose fair market value is $200, and it would cost $600 to repair the damage. Is this considered an accident?

a) No. The $1,000 threshold has not been met.

b) No. The fair market value is $200, which is under the $500 fair market value rule.

c) Yes. All accidents must be reported to the FAA.

Answer – b. The $500 threshold applies to the lesser amount of repair or replacement. In this case, the fair market value of $200 is the lesser amount. Reference: AC 107-2 (4.5)

237. To be classified as a sUAS accident under the FAA's definition, it must be reported to the FAA within 10 calendar days and

a) consist of a total loss of the sUAS no matter the value. A total loss signifies the serious nature of the accident and requires such classification.

b) consist of a serious injury, Level 3 or above (or any loss of consciousness), or damage to property, other than the sUAS, if the cost is greater than $500 to repair or replace the property (whichever is lower).

c) consist of a serious injury, Level 3 or above) and damage to property, other than the sUAS, if the cost is greater than $500 to repair or replace the property (whichever is lower).

Answer – b. Remember that meeting either of these thresholds, injury to persons or damage to property, is a requirement for reporting as an accident to the FAA FSDO or Regional Operations Center (ROC). Reference: AC 107-2 (4.5)

238. If a sUAS aircraft causes $200 worth of damage to property whose fair market value is $600, the accident

a) is required to be reported to the FAA because the repair cost is below $500.

b) is required to be reported to the FAA because the fair market value is $600, $100 over the limit.

c) is not required to be reported to the FAA because the repair cost is below $500.

Answer – c. If the cost of repair or replacement at fair market value is less than $500, it isn't required to be reported to the FAA FSDO or Regional Operations Center (ROC). Reference: AC 107-2 (4.5)

239. A sUAS accident must be reported within 10 calendar days by

a) telephone to the nearest jurisdictional FAA Flight Standards District Office (FSDO) or electronically to the FAA Regional Operations Center (ROC), instructions at; www.faa.gov/uas).

b) certified mail to the law enforcement agency that has jurisdiction over the flight mission area.

c) electronic means to the National Transportation Safety Board (NTSB) using online form A1A. A courtesy copy is required to be sent to the FAA Chief, National Operations Inspector in Washington, DC.

Answer – a. AC 107-2, CFR 107-2 and the FAA website give guidance regarding sUAS accident reporting in the form of "checklist" items such as; date, time, location, details of the event, injury, damage, etc. Reporting can be done through one of the numerous FAA field offices across the US or via an FAA Regional Operations Center (ROC). Reference: AC 107-2 (4.5.1)

A Guide to Better Decisions: Aeronautical Decision Making (ADM)

CRM and Task Management

240.   What is the definition of Crew Resource Management (CRM)?

a) Critical task management skills used by the Visual Observer (VO) that allows for advanced supervision of the Remote Pilot in Command (RPIC) and the Person Manipulating the Controls (PMC)

b) Effective use of the FAA CRM software that allows sUAS operators to schedule flight crews, track maintenance and report accidents (major and minor) to the appropriate FAA Flight Standards District Office (FSDO)

c) The effective use of all available resources for sUAS flight crew personnel to assure a safe and efficient operation, reducing error, avoiding stress and increasing efficiency.

Answer – c. Cockpit Resource Management (CRM) was first developed in the 1960's to help airline crews work better as a team rather than a dogmatic hierarchy with the Captain as the only decision maker. The crew works effectively by utilizing available resources such as; air traffic control, cabin crew, flight dispatch and technology for maximum efficiency. This same concept can be used to enhance the safety of the sUAS flight crew. CRM is under the larger umbrella of Aeronautical Decision Making (ADM).

241.   What are examples of the Remote Pilot (RP) exercising effective Crew Resource Management (CRM)?

a) Delegate task

b) Recognize Hazardous Attitudes as outlined in FAA publications

c) Establish effective team communications procedures

d) All of the above

Answer – d. The PRIC is at the top of the sUAS flight crew hierarchy, with that role comes with the responsibility of creating the atmosphere for effective CRM. Key areas of CRM are; delegation of a task, recognition of the hazardous attitudes (as stated in FAA publications), the establishment of effective communications procedures. Reference: Advisory Circular (AC) 107, AC 60-22, Pilots Handbook of Aeronautical Knowledge (PHAK) Ch. 1-4 and17-4.

5 Hazardous Attitudes

242. The 5 Hazardous Attitudes that the Remote Pilot (RP) and other crew members should be aware of and on the lookout for are;

a) Anti-Authority, Impulsivity, Assimilation, Macho and Resignation

b) Denial, Impulsivity, Invulnerability, Macho and Resignation

c) Anti-Authority, Impulsivity, Invulnerability, Macho and Resignation

Answer – c. The 5 Hazardous Attitudes are; Anti-Authority, Impulsivity, Invulnerability, Macho, and Resignation. Typical statements from those exhibiting the hazardous attitudes are; Anti-Authority – "Don't tell me," Impulsivity – "Do something quick," Invulnerability – "It won't happen to me," Macho – "I can do it," and Resignation – "What's the use." Reference: RMH, PHAK

243. A typical statement from someone exhibiting Anti-Authority hazardous attitude may state, "Don't tell me." The antidote to this statement would be:

a) Follow the rules they are usually right

b) Not so fast. Think first

c) It could happen to me

Answer – a. A person that is Anti-Authority doesn't like being told what to do, especially from those in authority. The appropriate mindset is to follow the rules because they are usually right. Reference: RMH, PHAK 17-5.

244. A typical statement from someone exhibiting Macho hazardous attitude may state, "I can do it." The antidote to this statement would be:

a) Follow the rules they are usually right

b) Not so fast. Think first

c) Taking chances is foolish

Answer – c. A person exhibiting a Macho hazardous attitude they feel they can do it by overcoming whatever obstacle is set in front of them. The correct thought pattern is to acknowledge that taking chances is foolish. Reference: PHAK 17-5.

245. A typical statement from someone exhibiting Impulsivity hazardous attitude may state, "Do something quickly." The antidote to this statement would be:

a) Follow the rules they are usually right

b) Not so fast. Think first

c) It could happen to me

Answer – b. A person displaying an Impulsive hazardous attitude will commonly act before thinking. The appropriate mindset is to think first and slow down. Reference: PHAK 17-5.

246. A typical statement from someone exhibiting Resignation hazardous attitude may state, "What's the use." The antidote to this statement would be:

a) Follow the rules they are usually right

b) Not so fast. Think first

c) I'm not helpless. I can make a difference

Answer – c. For some exhibiting the hazardous attitude of Resignation they feel there is no use in trying to correct a situation. The appropriate mindset is to realize a situation is not helpless, and a difference can be made. Reference: PHAK 17-5.

247. A typical statement from someone exhibiting Invulnerability hazardous attitude may state, "It won't happen to me." The antidote to this statement would be:

a) Follow the rules they are usually right

b) Not so fast. Think first

c) It could happen to me

Answer – c. A person with the Invulnerability hazardous attitude thinks something bad will not happen to them, but the correct thought pattern is to acknowledge it could happen. Reference: PHAK 17-5.

248. Identify the hazardous attitude or characteristic, a remote pilot displays while taking risks to impress others?

a) Impulsivity.

b) Invulnerability.

c) Macho.

Answer – c. Macho is the ego driven mode in which a person might try to impress others by showing off as the saying goes. References: Q18UAS SG

249. You are a remote pilot for a co-op energy service provider. You are to use your UA to inspect power lines in a remote area 15 hours away from your home office. After the drive, fatigue impacts your abilities to complete your assignment on time. Fatigue can be recognized

a) easily by an experienced pilot.

b) as being in an impaired state.

c) by an ability to overcome sleep deprivation.

Answer – b. The regulations state that a person can't operate a sUAS if he or she has a known medical deficiency, including lack of adequate sleep. Every pilot is required to "self-certify" they are in a physical and mental condition for flight. Studies have been done in which the performance of study subjects in a sleep deprived state is equated to a person under the influence of alcohol or drugs. References: CFR 107.17, Q19UAS SG

250. Safety is an important element for a remote pilot to consider prior to operating an unmanned aircraft system. To prevent the final "link" in the accident chain, a remote pilot must consider which methodology?

a) Crew Resource Management.

b) Safety Management System.

c) Risk Management.

Answer – a. The Accident Chain is a safety concept that states an accident isn't just one bad decision many multiple bad decisions. There are only so many bad decisions a pilot can make before an accident happens and the chain breaks. Crew Resource Management (CRM) is the effective use of all resources available to the flight crew to enhance the safety of flight. From the question, we gather that the scenario is an in-flight situation in which several bad decisions have already been made.

Safety Management System (SMS), answer b is incorrect because it is a formal system (comprised of four components) developed outside of the flight mission that encompasses the entire organization. Defined, SMS is the formal, top-down, organization-wide approach to managing safety risk and assuring the effectiveness of safety risk controls. It includes systematic procedures, practices, and policies for the management of safety risk. (FAA)

Answer c is incorrect because Risk Management techniques make up the components of the broader CRM while also having applications to an SMS.

While Safety Management Systems (SMS) and Risk Management are important and can be inter-related to CRM. CRM is the best answer due to the in-flight decision-making scenario set forth. References: FAA, RMH, Q20UAS SG

251.   When adapting Crew Resource Management (CRM) concepts to the operation of a small UA, CRM must be integrated into

a) the flight portion only.

b) all phases of the operation.

c) the communications only.

Answer – b. CRM was a crew concept developed by the airlines in the 1960's to help flight crews work better as a team. This team concept extends outside of the flight environment into the pre and post planning phase of flight. An example of this would be a sUAS post-flight debriefing to discuss issues of concern and along with positive outcomes. Remember, CRM falls under Aeronautical Decision Making (ADM). The answer still comes under CRM because pre-flight briefings and de-briefings are considered flight activities. References: RMH, PHAK, Q21UAS SG

252.   You have been hired as a remote pilot by a local TV news station to film breaking news with a small Unmanned Aircraft (UA). You expressed a safety concern and the station manager has instructed you to "fly first, ask questions later." What type of hazardous attitude does this attitude represent?

a) Machismo.

b) Invulnerability.

c) Impulsivity.

Answer – c. The station manager is expressing impulsiveness by wanting the job to be done immediately. sUAS flight crews have to be on guard for this type of behavior as to not be rushed into a bad decision. The antidote for impulsivity is to slow down and think first. If the station manager isn't willing to do this, it is up to the flight crew to protect the safety of the flight. This could call under Machismo and Invulnerability depending on how you look at it, but Impulsivity is the best answer. References: RMH, Q22UAS SG

253. A local TV station has hired a remote pilot to operate their small Unmanned Aircraft (UA) to cover breaking news stories. The Remote Pilot has had multiple near misses with obstacles on the ground, and two small UAS accidents. What would be a solution for the news station to improve their operating safety culture?

a) The news station should implement a policy of no more than five crashes/incidents within 6 months.

b) The news station does not need to make any changes; there are times that an accident is unavoidable.

c) The news station should recognize hazardous attitudes and situations to develop standard operating procedures that emphasize a safety culture.

Answer – c. The solution to this issue is to recognize any of the 5 Hazardous Attitudes (Anti-Authority, Impulsivity, Invulnerability, Macho and Resignation) that exist in the flight crew. Creating a standard of allowable accidents or ignoring the problem doesn't create a culture of learning from accidents/incidents and improving the safety culture of the organization. References: RMH, Q23UAS SG

Aviation Communication: ATC and sUAS Flight Crews

254. The correct method a manned aircraft would use to state 4,500 feet MSL to ATC is

a) "Four thousand five hundred."

b) "Four point five."

c) "Forty-five hundred feet MSL."

Answer – a. Most UAS thrives in the Class G and E environment, but it can be beneficial to know how communication (a.k.a. phraseology) accomplished for monitoring manned flight operations on an aviation band radio. Altitudes should be stated complete such as in the answer, "Four thousand five hundred." Reference: AIM

255. The correct method a manned pilot would use to state 10,500 feet MSL to ATC is

a) "Ten thousand, five hundred feet."

b) "Ten point five."

c) "One zero thousand, five hundred."

Answer – c. Once manned aircraft pilots are into the double-digit altitudes, the first two digits are stated individually, stating thousand last, then followed up normally with the hundreds of feet altitude. Reference: AIM

256. ATC advises a manned aircraft, "traffic 12 o'clock," the advisory is related to the aircraft

a) true course.

b) ground track.

c) magnetic heading.

Answer – b. On an aviation band radio, a UA flight crew may hear ATC gives traffic advisories (of other aircraft in the area) which are position reports relative to the ground track of the aircraft. Tip: Think about an air traffic controller behind a radar scope looking at aircraft "blips" on the screen. The only way he/she would be able to point out traffic is relative to the ground track, or blip, on the screen.

In manned aircraft pilots will point out traffic to other crew members or passengers by using a clock system in which, for example, 12 o'clock is directly in-front of the aircraft, 3 o'clock is off the right wing, 6 o'clock is directly behind and 9 o'clock is off the left wing (all numbers on the clock face are used too). Pilots can be more specific by stating the oncoming aircraft is "high," higher altitude than his/her aircraft, or "low," lower altitude than his/her aircraft. This same concept should be employed by the UA flight crew in some fashion to point out various hazards that crew members may need to point out to the team. Reference: AIM

257. An ATC radar facility issues the following advisory to a manned aircraft pilot flying on a heading of 090°, "traffic 3 o'clock, 2 miles, westbound." Where should the manned aircraft pilot look for the traffic?

a) East.

b) South.

c) West.

Answer – b. This manned aircraft example is a good way to practice this concept and employ it for the UA flight crew to point out in-flight hazards to the flight team. In the question the aircraft is heading 090°, which is East. If the traffic is at the pilots 3 o'clock position the pilot will look toward his/her right wing which is South. Tip: Feel free to draw it out on a sheet of paper to firmly think through and visualize the scenario. Reference: AIM

258. An ATC radar facility issues the following advisory to a manned aircraft pilot flying on a heading of 360°, "traffic 10 o'clock, 2 miles, southbound." Where should the pilot look for this traffic?

a) Northwest.

b) Northeast.

c) Southwest.

Answer – a. The pilot is flying on a heading of 360° or North. ATC states the traffic (or other aircraft in the area) is at 10 o'clock, which means the pilot would look left just ahead of the left aircraft wing, or Northwest. Reference: AIM

259. An ATC radar facility issues the following advisory to a manned aircraft pilot during a local flight; "traffic 2 o'clock, 5 miles, northbound." Where should the manned aircraft pilot look for this traffic?

a) Between directly ahead and 90° to the left.

b) Between directly behind and 90° to the right.

c) Between directly ahead and 90° to the right.

Answer – c. In this scenario the FAA doesn't give the direction of the aircraft, but you can still visualize or draw-out the scenario. In the aircraft straight ahead is the 12 o'clock position. If ATC gives a traffic advisory (other aircraft in the area) that another aircraft is at 2 o'clock, then the pilot will look to the right approximately 60°. Considering ATC doesn't know the exact angle of your aircraft with respect to wind (they only see aircraft as a "blip" on the screen) then we have to assume that the traffic can be anywhere between directly ahead to 90° to the right. Reference: AIM

260. An ATC radar facility issues the following advisory to a manned aircraft pilot flying North in a calm wind; "traffic 9 o'clock, 2 miles, southbound." Where should the manned aircraft pilot look for this traffic?

a) South.

b) North.

c) West.

Answer – c. The pilot flies North, or 360° (also 000°) and ATC states traffic (another aircraft) is at 9 o'clock. The pilot would look to the left, to the left wing, which is West. It is interesting that this question that they state the winds are calm, which means ATC will be able to give more accurate traffic advisories. Normally ATC radar scopes only show aircraft as "blips" and doesn't account for the effect of aircraft weathervaning into the wind. Reference: AIM

261. UA flight crew communication should include:

a) A discussion of the Remote Pilot (RP) that states all communication will be directed by the RP to preclude any other conversation that may distract the mission from going as planned and to allow for increased efficiency of the flight operation.

b) No communication standard is required on the flight mission field if the safety record of the flight crew is above 90%.

c) Mission pre-briefing and post-mission debriefing. Personal comm radios if the distance between crew members precludes unaided communication between each person in the crew, in addition to the adoption of a standard phraseology.

Answer – c. An effective flight crew will pre-brief the mission with a description of the mission and the roles of each crew member. This is followed by a debriefing that is an open discussion about the events that went well and those that can be learned from on future flights. If part of a multi-crew organization, it can be beneficial to pass this along to the entire organization as a positive learning experience. Standard phraseology (a standard dialog) should be adopted to ensure clear communication. If the mission covers a large distance comm radios should be employed (in addition to the monitoring of the local aviation band frequency if applicable). All crew members should have their own comm radio to prevent the relaying of important information. For example, if the VO is relaying a possible hazard the RP and Person Manipulating the Controls (PMC) at the Control Station (CS) should have a separate comm radio to prevent a breakdown in effective communication.

Flight Performance

262. What effect does high-density altitude have on the efficiency of a UA propeller?

a) Propeller efficiency is increased.

b) Propeller efficiency is decreased.

c) Density altitude does not affect propeller efficiency.

Answer – b. High-density altitude is the effect of not having many air molecules with which the Unmanned Aircraft (UA) propeller to utilize for thrust (and lift for aircraft with wings). The worst weather conditions for flight performance are high-density altitude issues such as; heavy aircraft weight, hot outside air temperatures, high altitude and high humidity (memory aid: heavy, hot, high, high). All of which have a limited air molecule density that the propeller can use for thrust. References: PHAK, Q36UAS SG

263. What is density altitude?

a) The height above the standard data plane.

b) The pressure altitude corrected for nonstandard temperature.

c) The altitude read directly from the altimeter.

Answer – b. Density altitude is one of the factors that can affect aircraft performance. Low-density altitude is good for aircraft because the air molecules are close together (high lift). For example, a low-density altitude day is associated with a cold winter, clear day at lower altitudes and low humidity.

Conversely, high density altitude results in poorer aircraft performance and, for example, are hot, humid days at higher altitude (elevation) locations. Memory aid: High density altitude is the most adverse on aircraft that are heavy (weight, lessens performance), hot (high temperatures) and high (high altitudes or elevation mission sites).

Since density altitude is a function of temperature creators of the concept created a Standard Day, which is 29.92 Barometric Pressure and 59° F (15°C). Density altitude is corrected for non-standard temperature. More on this: Manned aircraft pilots can set their altimeter barometric pressure to 29.92 while on the ground to find pressure altitude. Reference: PHAK

264.   What is pressure altitude?

a) The indicated altitude corrected for position and installation error in manned aircraft.

b) The altitude indicated when the barometric pressure scale is set to 29.92.

c) The indicated altitude corrected for nonstandard temperature and pressure.

Answer – b. Pressure altitude can be found by manned aircraft pilots by adjusting their altimeters barometric pressure to 29.92 (Standard Day barometric pressure) to find pressure altitude. Pressure altitude is used for Density Altitude calculations for aircraft performance. Reference: PHAK

265.   Under what condition is pressure altitude and density altitude the same value?

a) At sea level, when the temperature is 0 degrees F.

b) When the manned aircraft altimeter doesn't have an installation error.

c) At standard temperature.

Answer – c. Pressure altitude can be found by manned aircraft pilots by adjusting their altimeters barometric pressure to 29.92 (Standard Day barometric pressure) to find pressure altitude. Pressure altitude is used for Density Altitude calculations for aircraft performance. Reference: PHAK

266.   Under which conditions will pressure altitude to be equal to true altitude?

a) When the atmospheric pressure is 29.92 Hg.

b) When standard atmospheric conditions exist.

c) When indicated altitude is equal to the pressure altitude.

Answer – b. Pressure altitude can be found by manned aircraft pilots by adjusting their altimeters barometric pressure to 29.92 (Standard Day barometric pressure) to find pressure altitude. Pressure altitude is used for Density Altitude calculations for aircraft performance. True altitude is the height of the aircraft above Mean Sea Level (MSL), an average of Sea Level (since there are high and low tides). When standard atmospheric conditions exist, a.k.a. Standard Day, the barometric pressure is 29.92, temperature 59°F (15°C), true altitude and pressure altitude are equal. Interestingly, GPS based altimeters used in many UAS software systems is closer than altimeters to true altitude. Reference: PHAK

267.   How many satellites make up the Global Positioning System (GPS)?

a) 25

b) 22

c) 24

Answer – c. There are 2 dozen satellites that make up a full complement satellites. Word on the street is that there are a few spares but the FAA says, there are 24 so we will stick with this number. Reference: AIM

268.   What are the minimum number of Global Positioning System (GPS) satellites that observable by the user anywhere on earth?

a) 6

b) 5

c) 4

Answer – b. I was hoping we could say there were half a dozen satellites observable (because there are 2 dozen in total) by the user but it is one less. So much for a memory aid, but hopefully you can remember 5 is the answer. Reference: AIM

269. How many Global Positioning Systems (GPS) satellites are required to yield a three-dimensional position (latitude, longitude, and altitude) and time solution?

a) 5

b) 6

c) 4

Answer – c. Memory Aid: 4 satellites are required for 3D positioning (latitude, longitude, and altitude plus time solution). Having multiple satellites helps in triangulation calculations to get maximum information. Early in the satellite GPS program, it was difficult to get enough coverage for 3D position information. Bonus Memory Aid: There are 2 dozen satellites, there are 5 (one less than a half dozen) observable (for use) by anyone standing on Earth and it takes 4 satellites to have a 3-D position. Reference: AIM

270. The Global Positioning System (GPS) is

a) ground based.

b) satellite based.

c) antenna based.

Answer – b. Most people in recent times with GPS in cars and cell phones are very aware that GPS is satellite based. If not, SURPRISE! Reference: AIM, PHAK

Fit to Fly

271. Which is true regarding the presence of alcohol within the human body?

a) A small amount of alcohol increases vision acuity.

b) Consuming an equal amount of water will increase the destruction of alcohol and alleviate a hangover.

c) Judgment and decision-making abilities can be adversely affected by even small amounts of alcohol.

Answer – c. The message is not to trust all of the home remedies for curbing the effects of alcohol. Decision making is negatively affected and the only recovery solution is time. Remember the 8 hour "bottle to throttle" rule and 0.04 alcohol level rules. References: CFR 91.17, 91.19, Q12UAS SG

272. For a person holding or applying for a Remote Pilot: A conviction for the violation of any Federal or State statute relating to the growing, processing, manufacture, sale, disposition, transportation, or importation of narcotic drugs, marijuana, or depressant or stimulant drugs or substances is grounds for

a) being place on a probationary flight status for the period of 1 year from the time of conviction followed by a peer review. The peer review committee can add an additional 1 year of time to the probation with a maximum of 5 years.

b) denial of an application for an RP certificate with a small UAS rating for a period of up to 1 year after the date of final conviction. A person with an RP certificate can have his or her certificate suspended or revoked.

c) no adverse action by the FAA if the RP reports the incident within 1 year of the date of final conviction.

Answer – b. With any drug conviction outlined by regulation, it is required that it be reported to the FAA. In the modern world government computer systems are sharing more information across platforms. Not reporting a conviction can result in (up to) loss of the Remote Pilot certificate.

273. A sUAS crew cannot be under the influence of alcohol within how many hours prior to flight?

a) 12

b) 8

c) Zero. You can never be under the influence of alcohol and fly an aircraft.

Answer – c. Manned aircraft pilots are under the same regulation (91.17) regarding consumption of alcohol and time delay prior to flight. The memory aid is: 8 hours bottle to throttle. Cute, huh? One issue that has come up in situations is that while the pilot didn't consume alcohol within the 8-hour window they were still "under the influence" at the time of the flight which is also a violation; thus the answer is Zero. Reference: CFR 91.17

274. Who has the legal authority to require a sUAS flight crew member to submit to a test indicating the percentage by weight of alcohol in the blood?

a) Only the FAA has this authority.

b) Law enforcement officer.

c) National Transportation Safety Board (NTSB) board member.

Answer – b. The FAA realizes that in most situations local Law Enforcement will be the first on the scene with respect to alcohol and drone incidence. Therefore, the FAA has cooperation with the Law Enforcement community to handle blood alcohol testing. Law Enforcement also has the right to inspect a person's Remote Pilot certificate. CFR 107.59, 91.17

275. What is the maximum blood or breath alcohol concentration level considered to be illegal for UA flight?

a) 0.04 or greater.

b) 0.08 or greater.

c) There is no legal limit because it doesn't involve the use of a motor vehicle used on the public road system. However, flight crews are encouraged to adopt a consumption level dictated by their insurance policy parameters.

Answer – a. 0.04 or greater is the magic number. CFR 91.17

276. What statement best defines hypoxia?

a) A state of oxygen deficiency in the body.

b) An abnormal increase in the volume of air breathed.

c) A condition of gas bubble formation around the joints or muscles.

Answer – a. sUAS flight missions can take place in unique locations like high altitude mountain bases. High altitudes can deprive the human body of the necessary oxygen for mental and physical performance. Hypoxia is the effects of the lack of oxygen at high altitude on the human body. Reference: AIM

277. Which is not a type of hypoxia?

a) Histotoxic.

b) Hypoxic.

c) Hypertoxic.

Answer – c. Hypoxia is oxygen to the cells deficiency. Histotoxic hypoxia stems from tissue poisonings, such as that caused by cyanide, and certain other poisons like hydrogen sulfide.

Hypoxia (or being hypoxic) is a condition in which the body or a region of the body is deprived of adequate oxygen supply at the tissue level. Hypoxia can be part of the normal physiology, for example, during hypoventilation training or strenuous physical exercise.

Hypertoxic is the body's toxic condition of being affected by toxic poisoning. While histotoxic and hypertoxic are similar in the poisoning sources they are in different categories. References: PHAK, Wikipedia

278. Which of the following is a correct response to counteracting the feelings of hypoxia on a flight mission at elevations of 8,000 – 12,000' MSL?

a) Promptly go to a lower altitude and/or use supplemental oxygen. A better option is to plan to arrive at the location 1-3 days prior to acclimate to the higher altitude.

b) Avoid sudden inhalations.

c) FAA regulations require and acclimation plan of no more than 3 days on missions above 12,000' MSL. If missions are greater than 3 days, the crew must spend 3 days below 8,000' MSL before returning to the mission base camp.

Answer – a. High altitude mission requires pre-planning because the body may require acclimation to higher altitudes or suffer brain fog, headaches or worse. Even though Remote Pilots do not have to have a standard aviation medical the FAA does ask each pilot to "self-certify" he/she is in good health and able to safely perform flight duties prior to each flight. Reference: PHAK, https://www.princeton.edu/~oa/safety/altitude.html

279. sUAS flight crews should be able to overcome the symptoms and avoid future occurrences of hyperventilation by

a) closely monitoring the Control Station (CS) or flight controller to control the sUAS.

b) slowing the breathing rate, breathing into a bag, or talking aloud.

c) increasing the breathing rate in order to increase lung ventilation.

Answer – b. Flying a UA can be exciting to the point of being prone to hyperventilation. RP's should look for this in crew members instructing the afflicted to slow their breathing rate, talking aloud and/or breathing into a bag to normalize Carbon-Dioxide levels. Symptoms: lightheadedness, dizziness, blurred vision. Reference: PHAK, AIM

280. Rapid or extra deep breathing while using supplemental oxygen at a high altitude mission base can cause a condition known as

a) hyperventilation.

b) aerosinusitis.

c) aerotitis.

Answer – a. A UA can be perfect for high altitude mountain photography. Some people may use supplemental oxygen to sidestep the acclimation period. However, hyperventilation can still occur while using supplemental oxygen. Reference: PHAK, AIM

281. What are the symptoms of hyperventilation?

a) An Intense burst of energy.

b) Lightheadedness, dizziness, confusion, perspiration, fast breathing rate,

c) Swelling of the major joints such as; leg and arm that become more painful with movement.

Answer – b. Reference: PHAK

282. Since UA flying is primarily an outdoor activity in harsh, hot weather conditions there is the danger of heat exhaustion (**symptoms**: excessive thirst, weakness, headache and loss of consciousness, nausea and vomiting, muscle cramps, and dizziness). If you suspect heat exhaustion in yourself or flight crew member you should

a) not move to another location, say still and request medical assistance immediately.

b) reduce internal body temperature by; hydrating with hot water, moving to a sunny area and tightening clothing. Heat exhaustion can lead to the more serious heat stroke if not treated seriously.

c) reduce internal body temperature by; hydrating with cool water, moving to a shaded area, loosening clothing, using fans, etc. Heat exhaustion can lead to the more serious heat stroke if not treated seriously.

Answer – c. Reducing body temperature is the name of the game because you don't want heat exhaustion to turn into a heat stroke. Ref: FAA UAS ACS

283.   Heat Stroke is more serious than Heat Exhaustion?

a) True

b) False

Answer – a (True). Heat exhaustion can turn into heat stroke if the situation is not respected and treated seriously.

284.   When a stressful situation is encountered on a sUAS flight mission, an abnormal increase in the volume of air breathed in and out can cause a condition known as

a) hyperventilation.

b) aerosinustitis.

c) aerotitis.

Answer – a. If it is suspected you or a UA flight crew member is hyperventilating it is important to slow breathing, talk aloud and/or breath into a paper bag to normalize the carbon dioxide level. Reference: PHAK, AIM

285.   Which would most likely result in hyperventilation?

a) Emotional tension, anxiety, or fear.

b) The excessive consumption of alcohol.

c) An extremely slow rate of breathing and insufficient oxygen.

Answer – a. Hyperventilation is typically caused by emotional stressors with its associated tension, anxiety, and fear. Reference: PHAK, AIM

286. A pilot experiencing the effects of hyperventilation should be able to restore the proper carbon dioxide level in the body by

a) slowing the breathing rate, breathing into a paper bag, or talking aloud.

b) breathing spontaneously and deeply or gaining mental control of the situation.

c) increasing the breathing rate in order to increase lung ventilation.

Answer – a. Restoring the proper carbon dioxide levels can be done by these methods; slowing the breathing rate, breathing into a paper bag, or talking aloud. Reference: PHAK, AIM

287. The Person Manipulating the Controls (PMC) in a sUAS flight crew using First Person View (FPV), with a Beyond Visual Line of Sight (BVLOS) FAA waiver, can experience spatial disorientation if

a) they ignore the sensations of muscles and inner ear.

b) visual cues are limited or taken away, such as when entering low visibility conditions or with limited screen information (i.e. in an ascending turn, with primarily a sky view).

c) eyes are moved often in the process of cross-checking the FPV display.

Answer – b. When the information from the eyes doesn't match with what the body is experiencing can be a setup for spatial disorientation. Manned aircraft pilots that inadvertently fly into low visibility conditions, such as a cloud, and are not trained how to use the flight instruments to remain in control of the aircraft may incorrectly allow the sensation of their body to deceive what is actually happening to the aircraft. Similarly, divers in low visibility conditions may not know up from down and inadvertently swim the wrong way trying to find the surface. Reference: PHAK, AIM

288. The inability of a person to correctly determine his/her body position in space can happen to the UA pilot using First Person View (FPV) technology (requires FAA exemption approval). This reaction is the definition of

a) spatial disorientation.

b) hyperventilation.

c) hypoxia.

Answer – a. When the information from the eyes doesn't match with what the body is experiencing can be a setup for spatial disorientation. Manned aircraft pilots that inadvertently fly into low visibility conditions, such as a cloud, and are not trained how to use the flight instruments to remain in control of the aircraft may incorrectly allow the sensation of their body to deceive what is actually happening to the aircraft. Similarly, drivers in low visibility conditions may not know up from down and inadvertently swim the wrong way trying to find the surface. Reference: PHAK, AIM

289. A state of temporary confusion resulting from misleading information being sent to the brain by various sensory organs is defined as

a) spatial disorientation.

b) hyperventilation.

c) hypoxia.

Answer – a. When the information from the eyes doesn't match with what the body is experiencing can be a setup for spatial disorientation. Manned aircraft pilots that inadvertently fly into low visibility conditions, such as a cloud, and are not trained how to use the flight instruments to remain in control of the aircraft may incorrectly allow the sensation of their body to deceive what is actually happening to the aircraft. Similarly, divers in low visibility conditions may not know up from down and inadvertently swim the wrong way trying to find the surface. Reference: PHAK, A

290. Which technique should a sUAS flight crew use to scan for traffic in the flight operations area?

a) Systematically focus on 30° segments of the area for short intervals using the FAA endorsed the technique.

b) Concentrate on relative movement detected in the peripheral vision area.

c) Continuous sweeping off, the windshield from left to right.

Answer – a. Shifting focus in segments is the most effective way to scan for traffic and hazards in the flight operations area. Reference: PHAK, AIM

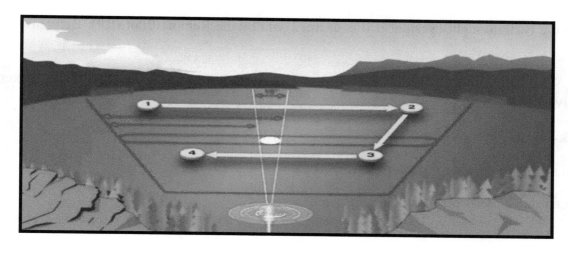

FAA sUAS crew scanning technique (source: FAA)

291. What effect does haze have on the ability to see potential aircraft traffic hazards during a flight mission?

a) Haze causes the eyes to focus at infinity.

b) The eyes tend to overwork in haze and do not detect relative movement easily.

c) All traffic or terrain features appear to be farther away than their actual distance.

Answer – c. Due to the reduction of visibility brought about by haze, traffic hazards may appear to be further away than their actual distance. Reference: PHAK, AIM

292. What preparation should a sUAS flight crew make to adapt their eyes for twilight or night flight (with a waiver)?

a) Wear sunglasses after sunset until ready for flight.

b) Avoid red lights at least 30 minutes before the flight.

c) Avoid bright white lights at least 30 minutes before the flight.

Answer – c. sUAS flight crews must use this technique to enhance their vision for low light conditions of twilight or night (with FAA waiver). The intensity of required anti-collision lights can be adjusted to preserve twilight/night eyesight if it is deemed safe by the Remote Pilot (RP). Reference: PHAK, AFH

293. What is the most effective way to use your eyes during night flight (with FAA exemption waiver)?

a) Look only as far away, dim lights.

b) Scan slowly to permit off-center viewing.

c) Concentrate directly on each object for a few seconds.

Answer – b. Night vision using the naked eye has the disadvantage of having a blind spot in the center of focus. Therefore, the guidance is to scan slowly using off-center viewing to see aircraft traffic, hazards and other objects of interest. Reference: PHAK, AIM, AFH

294. The best method to use when looking for other traffic and hazards at night (with FAA exemption waiver) is to

a) look to the side of the object and scan slowly.

b) scan the visual field very rapidly.

c) look to the side of the object and scan rapidly.

Answer – a. Night vision using the naked eye has the disadvantage of having a blind spot in the center of focus. Therefore, the guidance is to scan slowly using off-center viewing to see aircraft traffic, hazards and other objects of interest. Reference: PHAK, AFH

295. For sUAS flight crews that have aircraft or equipment that use combustion engines, for example; automobiles and generators for power, need to consider the possibility of carbon monoxide poisoning. The effect of this poisoning includes;

a) dizziness, blurred vision, and loss of muscle power.

b) sweating, increased breathing, and paleness.

c) motion sickness, tightness across the forehead, and drowsiness.

Answer – a. RPs should look for these symptoms in themselves and crews. However, carbon monoxide poisoning can cause a sense of euphoria and skewed thinking patterns. Reference: PHAK

296. Susceptibility to carbon monoxide poisoning increases as

a) altitude increases.

b) altitude decreases.

c) air pressure increases.

Answer – a. Carbon monoxide poisoning is a lack of oxygen reaching the body, so anything that enhances that condition, like an increase in altitude, will further enhance the effects of the poisoning. Reference: PHAK, AIM, AFH

297. Remember carbon monoxide is considered an odorless colorless gas making it potentially more hazardous. What is the correct response if carbon monoxide contamination is suspected?

a) Find a higher altitude location so the effects of CO would be decreased.

b) Take deep breaths so as to inhale more oxygen.

c) Move to an area with fresh air and consider cutting off combustion engines such as accessory power generators (automobiles and generators).

Answer – c. The objective is to get more fresh air and eliminate the source of carbon monoxide. Reference: PHAK, AFM

298. UA flight mission may take a flight crew to interesting destinations such as dive locations. Crews should be aware of the dangers of "the bends" (nitrogen comes out of solution in the joints) and return back home via an airline flight. What is the safest general guideline with respect of time delay for boarding an aircraft after diving?

a) Wait 3 days after diving prior to boarding an airline flight.

b) Wait 24 hours after diving prior to boarding an airline flight.

c) Wait 2 days after diving prior to boarding an airline flight.

Answer – b. There are many directives surrounding the time delay from diving to boarding and aircraft for flight based on depth of dive, etc. However, the best general guideline is to wait 24 hours after diving before boarding an aircraft.

Guide to Better Decisions: Aeronautical Decision Making (ADM)

299. Risk Management as part of the Aeronautical Decision Making (ADM) process relies on which features to reduce the risk associated with each flight?

a) Application of stress management and risk element procedures.

b) The mental process of analyzing all information in a particular situation and making a timely decision on what action to take.

c) Situational Awareness (SA), problem recognition, and good judgement.

Answer – c. ADM is the bigger concept that employees such tools as risk management, SA, problem recognition and good judgement. Reference: HFM, AC 60-22

300. What is it often called with a pilot pushes his or her capabilities and the aircraft's limits by trying to maintain visual contact with the terrain in low visibility and ceiling?

a) Scud running.

b) Mind set.

c) Peer pressure.

Answer – a. Scud running is a common terminology in the manned aircraft world pertaining to the push to complete a visual flight in low cloud ceiling and visibility environment. In this accident scenario, the pilot will fly lower and lower trying to maintain visual contact with the ground. Sadly, the visibility will reduce to point where obstacles can't be seen and a collision with the ground or an obstacle occurs. There are lessons that can be learned from the UA flight crew regarding, the pushing of a flight mission in limited visibility conditions. If the Person Manipulating the Controls (PMC) loses contact (sight) of the aircraft and loses control it can become a hazard to persons and property. Reference: PHAK, AFH, HFM

301. What often leads to disorientation, Loss Of Control (LOC), and collision with the ground or an obstacle while flying?

a) Continued flight into low visibility conditions.

b) Getting behind the aircraft.

c) Coriolis Effect

Answer – a. Manned aircraft pilots call this situation Scud Running in which the pilot attempts to maintain visual contact with the ground in deteriorating weather conditions (low visibility). This leads to a collision with the ground or an obstacle. For the UA pilot, this can mean an LOC event which can be hazardous to persons and property. Reference: PHAK, HFM, AC 60-22

302. What tool can be used to avoid forgetting a critical item when a sUAS flight crew relies on short and long term memory for repetitive task?

a) Checklist.

b) Situational Awareness (SA).

c) Flying outside of the envelope.

Answer – a. SA is more of an in-flight tool only, in which the crew is attentive to the entire flight environment for identifying issues and hazards. A checklist is a great tool for crews to before (preflight), during (in-flight) and after a flight (post-flight). Crews should come up with a method, such as a command and response method, for completing checklist in which one person reads the checklist item (command) and the other completes the task and verbal responses (response). Reference: PHAK, HFM, AC 60-22

303. Hazardous attitudes occur to every sUAS flight crew to some degree at some time. What are the 5 hazardous attitudes?

a) Antiauthority, impulsivity, macho, resignation, and invulnerability.

b) Poor situational awareness, snap judgements, indecision, and ambiguity.

c) Poor risk management, inattentiveness, indecision, ambiguity and lack of stress management.

Answer – a. Reference: PHAK, HFM, AC 60-22

304. In Aeronautical Decision Making (ADM) process, what is the first step in neutralizing a hazardous attitude?

a) Recognizing hazardous thoughts.

b) Recognizing the invulnerability of the situation.

c) Making a rational judgements.

Answer – a. The most difficult part in stopping hazardous attitudes is recognizing them, especially in ourselves. Reference: PHAK, HFM, AC 60-22

305. What is the antidote when a sUAS flight crew member has a hazardous attitude, such as "Antiauthority?"

a) Rules do not apply in this situation.

b) I know what I am doing.

c) Follow the rules.

Answer – c. Antidotes are responses a person uses to combat a hazardous attitude, such as with anti-authority one should follow the rules. Reference: PHAK, HFM, AC 60-22

306. What is the antidote when a sUAS flight crew member has a hazardous attitude, such as "Impulsivity?"

a) It could happen to me.

b) Dot it quickly to get it over with.

c) Not so fast, think first.

Answer – c. Antidotes are responses a person uses to combat a hazardous attitude, such as with impulsivity in which a person should concentrate on slowing down and thinking first. Reference: PHAK, HFM, AC 60-22

307. What is the antidote when a sUAS flight crew member has a hazardous attitude such as "Invulnerability?"

a) It will not happen to me.

b) It cannot be that bad.

c) It could happen to me.

Answer – c. Antidotes are responses a person uses to combat a hazardous attitude, such as with invulnerability the thought process should be that it could happen. Reference: PHAK, HFM, AC 60-22

308. What is the antidote when a sUAS flight crew member has a hazardous attitude such as "Macho?"

a) I can do it.

b) Taking chances is foolish.

c) Nothing will happen.

Answer – b. Antidotes are responses a person uses to combat a hazardous attitude, such as with acting macho in which the thought pattern should be that taking chances is foolish. Reference: PHAK, HFM, AC 60-22

309. What is the antidote when a sUAS flight crew member has a hazardous attitude such as "Resignation?"

a) What is the use?

b) Someone else is responsible.

c) I am not helpless.

Answer – c. Antidotes are responses a person uses to combat a hazardous attitude, such as with resignation in which a person should think that they are not helpless. Reference: PHAK, AC 60-22

310. Who is responsible for determining whether a sUAS flight crew member is fit for a flight mission?

a) The FAA.

b) An FAA AeroMedical Examiner (AME).

c) The sUAS flight crew, especially the Remote Pilot (RP) as the FAA certificate holder and highest in the flight crew hierarchy. However, the RP is additionally responsible for observing fitness for flight duty of his/her team while encouraging them to self-certify so they can safely complete the flight before each mission.

Answer – c. Each UA crew member should self-certify they can complete the flight mission safely. However, the RP should keep in mind they are the FAA certificate holder and therefore should look for fitness of flight in not only themselves but the entire flight crew. Crews should be educated on the self-certification concept to enhance safety of the flight mission. Reference: PHAK, HFM, AC 60-22

311. What is the one common factor which affects most preventable accidents?

a) Structural failure.

b) Mechanical malfunction.

c) Human error.

Answer – c. To err is pilot, most accidents that are reported in the media are blamed on "pilot error," a.k.a. human error. However, it is important to look beyond the flight crew for tools that can push safety forward. Reference: PHAK, HFM, AC 60-22

312. What antidotal phrase can help reverse the hazardous attitude of impulsivity?

a) Do it quickly to get it over with.

b) It could happen to me.

c) Not so fast, think first.

Answer – c. Antidotes are responses a person uses to combat a hazardous attitude, such as with impulsivity in which the thought pattern should be to slow down and think first. Reference: PHAK, HFM, AC 60-22

Rules of the Air

313. Figure 26, Area 2: While monitoring the Cooperstown CTAF you hear an aircraft announce that they are midfield left downwind to RWY 13. Where would the aircraft be relative to the runway?

a) The aircraft is East.

b) The aircraft is South.

c) The aircraft is West.

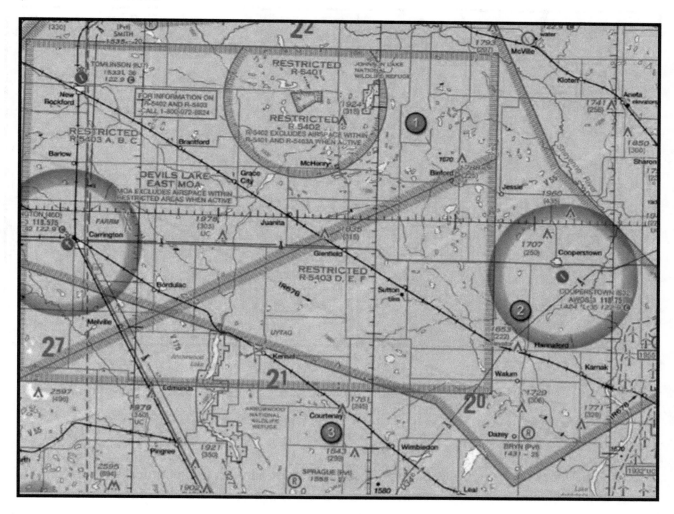

Figure 26 (Not to scale).

Answer – a. Using Figure 26 isn't necessary, but after cross-checking the location on the chart, we suggest drawing out the runways and related traffic pattern. Since the reporting pilot is landing Runway

(RW) 13, the pilot will be landing Southeast, or specifically 130 degrees. Standard traffic patterns are to the left, or alternatively stated, all turns are to the left once established in the traffic pattern (minus the entry point turn typically). The traffic pattern places the pilot on the East side of the airport. References: Q15UAS SG, FAA-CT-8080-2G.

314. Figure 22, Area 2: At Coeur D`Alene, which frequency should be used as a Common Traffic Advisory Frequency (CTAF) to monitor airport traffic?

a) 122.05 MHz.

b) 135.075 MHz.

c) 122.8 MHz

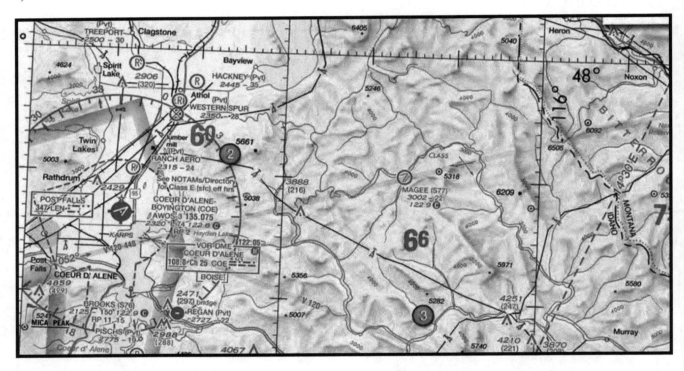

Figure 22 (Not to scale).

Answer – c. Under the name of the Coeur D' Alene Airport name is limited information such as (in order); airport identifier, AWOS weather observation frequency, elevation, lighting denoted by L, length of longest runway and Unicom frequency where pilots can give position reports on an open frequency. The last piece of information is the Unicom frequency 122.8. sUAS flight crews should monitor the frequency and notify the non-towered airport operator regarding the flight operation. References: Aeronautical Chart legend. Q24UAS SG, FAA-CT-8080-2G.

315. Figure 26, Area 4: You have been hired to inspect the tower under construction inside of the magenta dashed line outside of the Jamestown Regional (KJMS) Airport. What must you receive prior to flying your unmanned aircraft in this area?

a) Authorization from the military.

b) Notification of the airport operator be it city, county or contractor run.

c) Authorization from the National Park Service.

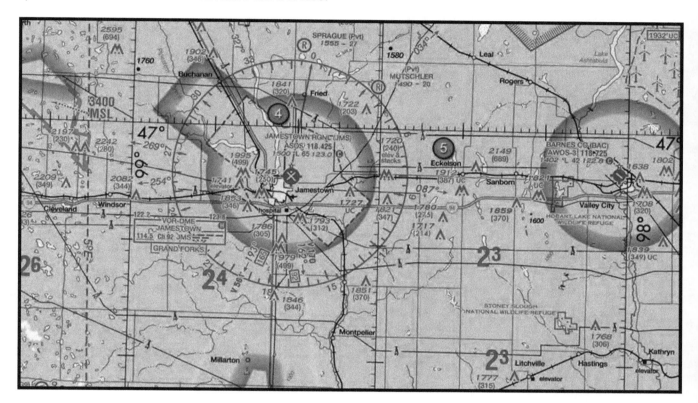

Figure 26, Area 4 (Not to scale).

Answer – b. The magenta dashed line around the KJMS Airport is Class E Airspace that extends to the ground instead of the 700' AGL if the dashed line didn't exist. The airport does not have an air traffic control tower (a.k.a. non-towered) but has instrument approaches into the airport in which they employ stricter weather requirements for manned aircraft to prevent midair collisions. Since KJMS is a non-towered airport a UA operator is required to notify the airport manager be it run by the city, county or contractor. References: Q25UAS SG, FAA-CT-8080-2G.

316. Figure 20, area 3: With ATC authorization, you are operating a small unmanned aircraft approximately 4 SM southeast of Elizabeth City Regional Airport (ECG). What hazard is indicated to be in that area?

a) High density military operations in the vicinity.

b) Unmarked balloon on a cable up to 3,008 feet AGL.

c) Unmarked balloon on a cable up to 3,008 feet MSL.

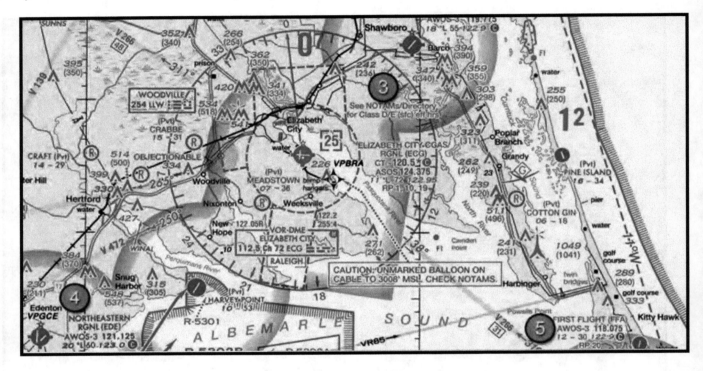

Figure 20, Area 3 (Not to scale).

Answer – c. If you look closely at the chart approximately 4 SM Southeast of KECG you can see a symbol with an arrow and associated information box. The text states there is a cable on a line extending to 3008' MSL. Useless Knowledge: This balloon on a cable is associated with a major unmanned balloon manufacturer that produces very large balloons used for security purposes. References: Q26UAS SG, FAA-CT-8080-2G.

317.   The most comprehensive information on a given airport is provided by

a) the Chart Supplements U.S. (formerly Airport Facility Directory).

b) Notices to Airmen (NOTAMS).

c) Terminal Area Chart (TAC).

Answer – a. The Chart Supplements U.S. is the most comprehensive resource for airport information. Formerly known as the Airport Facility Directory, or affectionately called the "green book" can now be found via electronic format. References: Q27UAS SG

318.   Which technique should a remote pilot use to scan for traffic? A remote pilot should

a) systematically focus on different segments of the sky for short intervals at 30° intervals beginning with the furthest distance in the field of view, sweeping left to right then shifting closer using a box pattern going from right to left.

b) concentrate on relative movement detected in the peripheral vision area.

c) continuously scan the sky from right to left as quickly as possible to take in the entire field of view to ensure all traffic and hazards are identified and sorted by the cognitive area of the brain that subconsciously goes back to the hazard if indeed a threat.

Answer – a. To harness the maximum performance from the eyes for scanning for traffic the recommendation is to scan small segments of the sky for a short duration, then scan another segment beginning with the farthest distance to the left first. Next, scanning right then focus closer then scan back to the left using 30° segments of the area. Peripheral vision (answer b), or off center viewing, is used as a technique for night flight operations, which is only allowed through an FAA approved waiver. Answer c is incorrect because rapid scanning is prone to not seeing aircraft in the area that may be a hazard. References: PHAK, Q29UAS SG

319. According to CFR part 107, what is required to operate a small UA within 30 minutes after official sunset?

a) Use of anti-collision lights.

b) Must be operated in a rural area.

c) Use of a transponder.

Answer – a. A sUAS can operate 30 minutes after official sunset (exceptions for Alaska) if it is equipped with anti-collision lights that are visible within 3 Statute Miles (SM). The Remote Pilot (RP) can reduce the intensity of the lights if it is determined the flight operation can be made safely considering the flight conditions. References: CFR 107.29, Q34UAS SG

320. An airport rotating beacon operated during daylight hours indicates

a) there are obstructions on the airport.

b) the weather at the airport located in Class D airspace is below basic VFR weather minimums.

c) the Air Traffic Control Tower is not in operation.

Answer – b. Airport officials will turn on the Rotating Beacon (an alternating green and white light located on a tower at standard civilian airports) when the weather is below Visual Flight Rules with ceilings below 1000' AGL and visibility less than 3 SM visibility. This is a good reference to the UA for weather conditions if in the airport area. UA pilots are required to have "no less than 3 SM" visibility and stay "500 feet below a cloud" and "2000 feet horizontally from the cloud." Reference: CFR, AIM, CFR 107.51

321. A lighted heliport can be identified by a

a) green, yellow, and white rotating beacon.

b) flashing yellow light.

c) blue lighted square landing area.

Answer – a. UA's and helicopters and stand the most chance of a midair collision due to the low flight paths of the respective aircraft. A green, yellow and white rotating beacon can warn a UA flight crew there are helicopter operations in the area. Reference: AIM

322. A military air station can be identified by a rotating beacon that has the following colors of lights:

a) white and green alternating flashes.

b) two quick, white flashes between green flashes.

c) green, yellow, and white flashes.

Answer – b. If a UA crew notices they are in the vicinity of a military airport (two quick white flashes between green flashes) while on the mission site (or near) the RP should investigate any airspace conflict prior to proceeding. Reference: AIM

323. Which is the correct traffic pattern departure manned aircraft use at a non-air traffic controlled airport?

a) Departures can be in any direction consistent with safety, outside of the airport boundary.

b) All departure turns will be to the left.

c) Departures should only be a maximum performance right climbing turn to cruise altitude.

Answer – c. The AIM recommends that departures from a runway depart straight ahead to traffic pattern altitude (800' – 1,000' AGL) prior to making a right turn (see diagram below). Otherwise, a left departure turn can take place directly after take-off. However, the AIM contains recommended procedures and pilots can elect to depart to the right if safety isn't compromised. Typically, pilots will gauge this on the traffic volume at the departure airport. Reference: PHAK, CFR 91.127

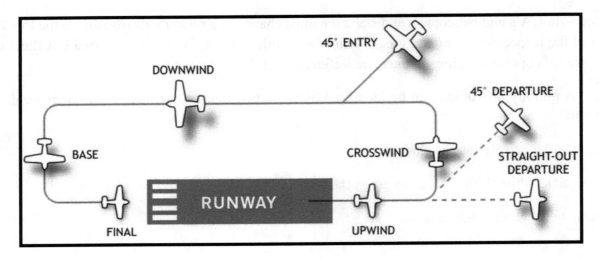

Standard Traffic Pattern (source: wikipedia)

324. What is the recommended traffic pattern entry procedure manned aircraft use at a non-towered airport?

a) 45 degrees to the base leg just below traffic pattern altitude.

b) enter 45 degrees at the midpoint of the downwind leg at traffic pattern altitude.

c) cross directly over the airport at traffic pattern altitude and join the downwind leg.

Answer – b. A non-towered airport means a non-air traffic control tower airport. The standard arrival procedure at this type airport (unless otherwise stated) is to make all turns to the left while in the traffic pattern (see graphic below). Entries into the pattern consist of a 45-degree entry (right turn) at the mid-point of the downwind leg. Reference: AIM, PHAK

325. The most effective method for scanning for other aircraft for collision avoidance is to use

a) regularly spaced concentration on the 3, 9, and 12 o'clock positions.

b) a series of short, regularly spaced eye movements to search each 10-degree sector.

c) peripheral vision by scanning small sectors and utilizing off-center viewing.

Answer b. Scanning sectors, is the best method of effectively scanning for (aircraft) traffic and other hazards. Reference: PHAK

326. Prior to starting each maneuver, UA crews should

a) check, altitude, airspeed, and heading indications.

b) visually scan the entire area for collision hazards.

c) announce their intentions on the nearest CTAF.

Answer – b. Any flight or maneuver should be started by carefully scanning the area for hazards. Reference: PHAK

327. Most midair collision accidents with manned aircraft occur during

a) hazy days.

b) clear days.

c) cloudy nights.

Answer – b. This is one of the strangest facts in aviation and feels it will remain the same in the UA community. Call it a false sense of security or something else we should always be vigilant but especially on clear days. Reference: PHAK

328. The definition of nighttime is

a) sunset to sunrise.

b) 1 hour after sunset to 1 hour before sunrise.

c) the time between the end of evening civil twilight and the beginning of morning civil twilight.

Answer – c. Let us begin with defining day. The day begins with morning civil twilight, and then official sunrise occurs. Late in the day official sunset happens followed by evening twilight. The end of evening twilight ends the day. The times in-between the described times is at night. Reference: CFR 107. 29

329. For manned aircraft: Except when necessary for takeoff and landing an aircraft may not be operated closer than what distance from any person, vessel, vehicle, or structure?

a) 500 feet.

b) 700 feet.

c) 1,000 feet.

Answer – a. The good news is that there is a built-in safety cushion of 100' between the maximum UA can fly (400') and the minimum manned aircraft can fly (500') with the exception of take-off and landing. Reference: CFR 91.119

330. What type of aircraft does a sUAS have priority with respect to right-way-rules?

a) Gliders and balloons.

b) Manned aircraft.

c) No other aircraft. A sUAS must yield the right-of-way to all other aircraft, airborne vehicles, and launch and reentry vehicles.

Answer – c. UA has to yield to all other aircraft. Until the advent of UAS regulations stated an aircraft in an emergency has right-of-way over all other aircraft. Reference: CFR 107.37

331. How close can a sUAS operate to another aircraft?

a) A 500-foot radius is required buffer between aircraft unless allowed by ATC.

b) Not so close to another aircraft as to create a collision hazard.

c) As close as allowed in formation flight, but requires prior coordination between crews.

Answer – b. The bottom line is not so close to creating a collision hazard. Reference: 107.37

332. What is the definition of a UA yielding the right-of-way to all other aircraft, airborne vehicles, and launch and reentry vehicles?

a) A sUAS may not pass over, under, or ahead of it (all other aircraft) unless well clear.

b) A sUAS must contact the aircraft via aviation band radio and receive approval from the other aircraft's Pilot in Command (PIC).

c) A sUAS must fly to the right of other aircraft using the prescribed right-of-way rules in CFR Part 91.

Answer – a. Many UA RPs will suspend flight operations if manned aircraft are seen or heard in the flight mission area. If you can't pass over, under or ahead the best policy is to temporarily suspend operations. Reference: CFR 107.37

333. When can a sUAS flight crew fly over human beings?

a) Regulations specifically prohibit flying over human beings.

b) The flight is only approved for people with applications made to the National Transportation Safety Board (NTSB) Aviation Safety Reporting System (ASRS).

c) Only when the people (human beings) are directly participating in the operation or under a covered structure or stationary vehicle.

Answer – c. Protection of people and property is important during UA flight operations. There are many examples online of consumer drones flying over large groups of people. I remember a fitting safety message from an Industrial Safety class in college in which the professor stated, "a 10-pound weight dropped from 10 feet can kill a person." Wise words. A UA operator can apply for waiver to fly over people. CFR 107.39

334. Under what conditions can a sUAS carry hazardous material?

a) Only if approved by formal agreement letter by the Environmental Protection Agency (EPA).

b) sUAS may not carry hazardous materials at any time unless approved via the waiver process listed in CFR 107.205, "List of regulations subject to waiver."

c) sUAS may not carry hazardous materials at any time. Items listed via the waiver process CFR 107.205, "List of regulations subject to waiver," which do not include hazardous materials.

Answer – c. No hazardous materials can be carried, common sense unless you count the lithium battery pack (don't take that last part seriously). Reference CFR 107.200, 107.205

335. If an in-flight emergency requires immediate action, the Remote Pilot (RP) may

a) deviate from any part of CFR Part 107 to the extent necessary to meet the emergency. The FAA Administrator may request a written report of the deviation.

b) not break any regulation covered in CFR Part 107 because it may endanger other flight operations in the area.

c) deviate from any part of CFR Part 107 to the extent necessary to meet the emergency. The National Transportation Safety Board (NTSB) Board Member will require a written report of the deviation.

Answer – a. This is an interesting question because UA's are required to give way to all other aircraft in the case of an emergency. According to CFR 107.21, a UA pilot can deviate from any regulation to meet the emergency. I would interpret this to mean common sense is applied to an emergency since UA are not manned (sorry to be redundant).

Pre-flight and Maintenance

Airworthiness

336. Manned aircraft are required to have an Airworthiness Certificate and so do sUAS aircraft.

a) True

b) False

Answer – b. Airworthy is a term to describe an aircraft that is deemed safe for flight. To remember think; *Worthy* of the *Air*. FAA certificated manned aircraft are required to have an airworthiness certificate, but sUAS do not. Since sUAS are much smaller aircraft the FAA has agreed that no manufacturing standard is necessary. However, this does not cancel the requirement for sUAS operators to thoroughly check the aircraft, called a pre-flight for a safe operating condition before flight. Reference 14 CFR 107.15 and Advisory Circular (AC) 107.

337. A preflight check is required;

a) On an annual basis as part of the annual inspection only and must be documented in the logbook

b) The night prior to planned flight the preceding day

c) Immediately prior to a flight to ensure the aircraft is in a safe condition

Answer – c. A pre-flight is a physical inspection of the sUAS prior to each flight to ensure it is safe for flight, a.k.a. airworthy.

338. Under what condition should the operator of a small UA establish scheduled maintenance protocol?

a) When the manufacturer does not provide a maintenance schedule.

b) UAS does not need a required maintenance schedule.

c) When the FAA requires you to, following an accident.

Answer – a. While not a certificated aircraft the FAA requires by regulation as a way to track and log via hard copy or electronic means sUAS maintenance. In the absence of a manufacturer maintenance schedule, one needs to be developed for submission to the FAA upon request. References: CFR 107.7, Q16UAS SG

339. According to 14 CFR part 107, the responsibility to inspect the small UAS to ensure it is in a safe operating condition rests with the

a) Remote Pilot.

b) visual observer.

c) owner of the small UAS.

Answer – a. The Remote Pilot (RP) is at the top of the sUAS flight crew hierarchy (followed by the Person Manipulating the Controls and Visual Observer) and responsible for ensuring the aircraft is in a condition for safe operation through inspection. References: CFR 107.15, Q17UAS SG

340. What should the Remote Pilot (RP) know about any safety recalls and notices received from the manufacturer?

a) For informational purposes only.

b) They are mandatory.

c) They are voluntary.

Answer – b. For commercial UAS operations, any safety recalls and notices are required to be complied with on the schedule stated by the manufacturer. Reference: CFR 107.15

341. May a Remote Pilot (RP) operate an aircraft that has not complied with a manufacturer's safety recalls and notices?

a) Yes, but only for one hour of flight time per calendar day.

b) Yes, manufacturer safety recall is voluntary.

c) Yes, if the manufacturer states in the safety recall and notices a provision for safe operation until compliance.

Answer – c. If the nature of the safety recall is serious enough, a manufacturer may direct the owner/operator to immediately suspend operations of the unit. In less serious situations the manufacturer may state it should be corrected within a certain time frame or hours of usage. Reference: CFR 107.15

342.   Who may perform maintenance on a sUAS aircraft?

a) Remote Pilot (RP) or a qualified person designated by the RP, such as the manufacturer and/or general RC repair personnel.

b) Only an FAA certificated Airframe and Powerplant maintenance technician.

c) Maintenance on a sUAS the aircraft is restricted only to the manufacturer. Parts and accessories can be swapped out to reduce the down time of the aircraft for flight missions.

Answer – a. sUAS aircraft have a lot more leeway when it comes to the maintenance on their aircraft. Manned aircraft pilots are required to use FAA certificated Aircraft Maintenance Technicians (AMTs) (Airframe and Powerplant mechanics) to accomplish maintenance. It makes sense that the RP can accomplish the maintenance task or find a qualified person to do so. Don't forget to log any maintenance on the aircraft. Reference: CFR 107.15

343.   For safety reasons the Remote Pilot (RP) should accomplish and document all maintenance as directed by the manufacturer and because the FAA Administrator

a) requires the RPIC to attend an annual maintenance safety system compliance conference and volunteer all maintenance data.

b) may make any test or inspection of the sUAS and related documentation.

c) wants to catalog maintenance of all sUAS to develop official Airworthiness Directives for sUAS aircraft.

Answer – b. Documentation of sUAS maintenance is important in case of FAA audit. The term "FAA Administrator" simply means someone (like an FAA Maintenance Inspector) is working under the authorization of the FAA Administrator. The Administrator would be one busy person if he/she had to do all of the items referenced in the regulations. Reference: 107.7

344. A UA flight crew just received an aircraft back from the manufacturer after a critical repair had been made to the battery system to prevent the system for catching on fire while under heavy electrical loads. The sUAS was shipped priority overnight for a flight mission that same morning. What action should the Remote Pilot (RP) take prior to the mission flight?

a) Test fly the sUAS in a safe area, under controlled conditions, prior to taking to the job site unless the job site is rural enough for such an operation. Additionally, the test flight should be documented in the maintenance records for FAA review if the need arises.

b) The work by the manufacturer will suffice because the FAA requires factory testing per CFR 107. A record of the maintenance, with a factory inspection seal, may be included in the maintenance logs. Additionally, a waiver must be signed by the property owner of the test flight in addition to the mission.

c) Due to the maintenance accomplished by the manufacturer being minor in nature, it is acceptable to fly the sUAS for the mission flight without any further action. No notation or documentation of manufacturer maintenance is required because the FAA can contact the manufacturer for a record of any maintenance accomplished at the factory for a specific sUAS

Answer – a. Test flying after maintenance is an important task for the safety of the flight mission site and personnel. The test flight should be in a safe area and documented in the maintenance records. Test flying is a small amount of insurance against a liability claim. Reference: CFR 107.7, 107.15

345. Where are the operating limitations of a UA found?

a) On the Airworthiness Certificate.

b) In the manufacturers' owner's manual for the sUAS, recall/safety communication, markings, and placards, or any combination thereof.

c) In the aircraft maintenance logbook(s).

Answer –b. The manufacturer is key when it comes to operating limitations. It is encouraged to make the owner's manual part of a larger operation manual that contains standard operating procedures, etc. CFR 107.15

346.   Preflight of a sUAS prior to flight must include;

a) local weather conditions, local airspace and flight restrictions, location of persons and property on the surface, ground hazards, personnel briefing, general mechanical, control link between ground stations and UA, power requirements, and securing of accessories/objects.

b) national weather conditions overview, all airspace and flight restrictions for the state of operation, location of property on the surface, animal hazards, general mechanical, a control link between ground stations and UA, and securing of accessories/objects.

c) A quick pre-flight, operational check, discussion with the property owner about area flight restrictions, securing of non-lashed items in the flight operations area.

Answer – a. Flight preparation typically begins well before arriving on the flight mission site. A pre-check of the mission site with respect to weather, airspace and hazards are followed by an on-site crew briefing. A physical pre-flight check of the aircraft and Control Station components is required just prior to the motor start. Reference: CFR 107.51

Aerodynamics

347.   When operating an unmanned airplane, the remote pilot should consider that the load factor on the wings may be increased anytime

a) the CG is shifted rearward to the aft CG limit.

b) the airplane is subjected to maneuver other than straight and level flight.

c) the gross weight is reduced.

Answer – b. When a conventional wing aircraft is turned, there is less available wing area (with respect to the earth, to provide lift and thus load factor is increased. a) An aft CG does not affect load factor. c) load factor will increase with an increase in gross weight, not a decrease in weight. Reference: PHAK, Q9UAS SG.

348. A stall occurs when the smooth airflow over the unmanned airplane's wing is disrupted, and the lift degenerates rapidly. This is caused when the wing

a) exceeds the maximum speed.

b) exceeds maximum allowable operating weight.

c) exceeds its critical angle of attack.

Answer – c. An aircraft wing lift depends on smooth airflow over and under the wing. When, for example, the aircraft climbs very steeply (approximately 15-17 degrees from the original straight and level flight) the airflow is disrupted. This disruption is called the critical angle of attack, an imaginary line dissecting the wing and the oncoming wind, called relative wind, causes a stall. The stall may be witnessed as a buffeting of the aircraft as it struggles to fly. The buffeting may increase to the point in which the aircraft's angle of attack decreases rapidly or falls off to the side into a spin. Reference: PHAK, Q10UAS SG

349. Figure 2: If an unmanned airplane weighs 33 pounds, what approximate weight would the airplane structure be required to support during a 30° banked turn while maintaining altitude?

a) 34 pounds.

b) 47 pounds.

c) 38 pounds.

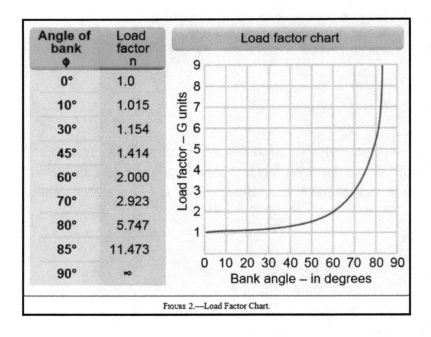

Figure 2.—Load Factor Chart.

Answer - c. Finding the net G force on the aircraft using Figure 2 requires multiplying the 1 G weight of 33 lbs. by the G force (Load Factor) of 1.154 (30 degrees of bank) which equals 38.082 lbs. (closest answer). While the graphic on the right side of Figure 2 gives a rough idea of G force (Load Factor) with an angle of bank use the left-hand side to plug in the correct multiplier depending on angle of bank. References: PHAK, FAA-CT-8080-2G, Q11UAS SG

350. What is the one purpose of wing flaps on a fixed wing UA?

a) To enable the pilot to make steeper approaches to a landing without increasing the airspeed.

b) To relieve the pilot of maintaining continuous pressure on the controls.

c) To decrease wing area to vary the lift.

Answer – a. Flaps are rare in most sUAS, but a good aerodynamic concept to understand. Flaps increase the camber of the wing, increasing the lifting surface. However, there is a break-even point because the polar opposite of lift is a drag (stated another way; with lift, you will always and drag). When flaps are lowered to $10°$, for example, the aircraft has more lift than drag. Lower the flap to $35°$ and the aircraft will have more drag than lift. The key is using flaps to enhance the characteristics the pilot desires. Taking off on a short runway the pilot may select $10°$ of flaps to increase lift and make the aircraft take-off in a shorter distance. Landing the pilot can use full flaps to approach a runway steeply and land in a shorter distance due to reduced landing speeds. Stall speed is also less as flaps are deployed. Reference: PHAK

351. One of the main functions of flaps during approach and landing on a fixed wing UA is to

a) decrease the angle of descent without increasing the airspeed.

b) permit a touchdown at a higher than indicated airspeed.

c) increase the angle of descent without increasing the airspeed.

Answer – c. Flaps are rare in most sUAS, but a good aerodynamic concept to understand. Flaps increase the camber of the wing, increasing the lifting surface. However, there is a break-even point because the polar opposite of lift is drag (stated another way; with lift you will always and drag). When flaps are lowered to $10°$, for example, the aircraft has more lift than drag. Lower the flap to $35°$ and the aircraft will have more drag than lift. The key is using flaps to enhance the characteristics the pilot desires.

Taking off on a short runway the pilot may select 10° of flaps to increase lift and make the aircraft take-off in a shorter distance. Landing the pilot can use full flaps to approach a runway steeply and land in a shorter distance due to reduced landing speeds. Stall speed is also less as flaps are deployed. Reference: PHAK

352. What is the purpose of the rudder on the fixed wing UA aircraft?

a) To control yaw.

b) To control the over banking tendency.

c) To control roll.

Answer – a. We are talking about 3 axes controlled aircraft that employ ailerons, elevator, and rudder. As the aileron is actuated to turn the down aileron creates lift and drag. The drag causes the wing to be held back, so to speak, creating a yawing motion. The rudder is designed to overcome the drag called adverse (bad) drag. Reference: PHAK

353. Which is not a primary flight control surface on a fixed wing UA?

a) Flaps.

b) Stabilator.

c) Ailerons.

Answer – a. Ailerons turn the aircraft via the aircraft's longitudinal axis (what I like to call the armrest axis because it runs the length of the fuselage). The stabilator, an elevator like tail surface in which the whole tail surface moves instead of a portion that is attached to the fixed horizontal stabilizer. The stabilator/elevator moves the aircraft around the lateral (wingtip to wingtip) axis of the aircraft. Flaps are an aerodynamic device used to create different proportions of lift and drag depending on the amount uses, such as 10, 20 or 30 degrees. This isn't a primary flight control. Reference: PHAK

354. The elevator controls movement around which axis on a fixed wing UA?

a) Longitudinal.

b) Lateral.

c) Vertical.

Answer – b. The elevator controls the pitch of the aircraft and moves around the lateral axis (wingtip to wingtip) axis of the aircraft. Reference: PHAK

355. Which statement is true concerning primary flight controls of a fixed wing UA?

a) The effectiveness of each control surface increases with speed because there is more flow over them.

b) Only when all three primary flight controls move in sequence to the airflow and pressure distribution change over and around the airfoil.

c) Primary flight controls include ailerons, rudder, elevator, and trim systems.

Answer – a. Under low-speed conditions, it will take more control input to get a response due to reduced airflow. Reference: PHAK

356. Which of the following is true concerning flaps on a fixed wing UA?

a) Flaps are attached to the leading edge of the wing and are used to increase wing lift.

b) Flaps allow an increase in the angle of descent without increasing airspeed.

c) Flaps are high drag devices deployed from the wings to reduce lift.

Answer – b. Flaps are rare in most sUAS, but a good aerodynamic concept to understand. Flaps increase the camber of the wing, increasing the lifting surface. However, there is a break-even point because the polar opposite of lift is drag (stated another way; with lift you will always and drag). When flaps are lowered to 10°, for example, the aircraft has more lift than drag. Lower the flap to 35° and the aircraft will have more drag than lift. The key is using flaps to enhance the characteristics the pilot desires. Taking off on a short runway the pilot may select $10°$ of flaps to increase lift and make the aircraft take-off in a shorter distance. Landing the pilot can use full flaps to approach a runway steeply and land in a shorter distance due to reduced landing speeds. Stall speed is also less as flaps are deployed. Reference: PHAK

357.   Which devices act as a secondary flight control on a fixed wing UA?

a) Spoilers.

b) Ailerons.

c) Stabilators.

Answer – a. Ailerons control, banking (for turns) flight around the longitudinal axis (armrest axis as I like to call them) of the aircraft. Stabilators, like elevators, control pitch around the lateral (wingtip to wingtip) axes of the aircraft. Reference: PHAK

358.   Trim systems are designed to do what on UA aircraft?

a) They relieve the pilot of the need to maintain constant back pressure on the flight controls.

b) They are used during the approach and landing to increase wing lift.

c) They move in the opposite direction from one another to control roll.

Answer – a. More common with UA but some may employ the use of trim systems (flight or ground adjustable) to help the pilot from having to hold constant elevator back pressure, for example, to prevent the aircraft losing altitude when desiring straight and level flight. Trim systems can also be employed on the rudder and ailerons as well. Reference: PHAK

359.   The four forces acting on a fixed wing UA aircraft are

a) lift, weight, thrust, and drag.

b) lift, weight, gravity, and thrust.

c) lift, gravity, power, and friction.

Answer – a. Lift acts in an upward force countered by weight (or gravity). Thrust from the propeller, is countered by drag. Unless in straight and level unaccelerated flight the forces are unequal in force. Reference: PHAK

360. When are the four forces that act on a fixed wing UA in equilibrium?

a) During un-accelerated level flight.

b) When the aircraft is accelerating.

c) When the aircraft is at rest on the ground.

Answer – a. In straight and level un-accelerated flight the four forces are in equilibrium. Lift equals weight (or gravity) and thrust equals drag. Reference: PHAK

361. What is the relationship of lift, drag, thrust, and weight when the fixed wing UA is in straight-and-level flight?

a) Lift equals weight and thrust equals drag.

b) Lift, drag, and weight equal thrust.

c) Lift and weight equal thrust and drag.

Answer – a. With respect to the four forces, lift opposes weight (or gravity) and Thrust opposes drag. When in straight and level unaccelerated flight the four forces are in equilibrium. Reference: PHAK

362. Which statement relates to Bernoulli's Principle?

a) For every action, there is an opposite and equal reaction.

b) An additional upward force is generated, as the lower surface of the wing deflects air downward.

c) Air traveling faster over the curved upper surface of an airfoil causes lower pressure on the top surface.

Answer – c. Bernoulli's Principal is one of the concepts that relate to the creation of lift. The low pressure on top of the wing lifts the wing into the air. You can get a better idea of the principal with a carburetor, as the air passes the constriction airflow speeds up, creating a low pressure. Each side of the carburetor represents an aircraft wing with the constrictions being made on the curved side of the wing. Reference: PHAK

363. The term Angle of Attack (AOA) on a fixed wing UA is defined as the angle between the

a) chord line of the wing and the relative wind.

b) airplane's longitudinal axis and that of the air striking the airfoil.

c) airplane's center line and the relative wind.

Answer – a. The chord line of a wing is the imaginary line that dissects the wing in the middle when looking at a cross-section of the wing. The relative wind is the oncoming wind and the angle between the chord line and wind is AOA. Reference: PHAK

364. The term "angle of attack" is defined as the angle

a) between the wing chord line and the relative wind.

b) between the aircraft's climb angle and the horizon.

c) formed by the longitudinal axis of the aircraft and the chord line of the wing.

Answer – a. The chord line of a wing is the imaginary line that dissects the wing in the middle when looking at a cross-section of the wing. The relative wind is the oncoming wind and the angle between the chord line and the wind is AOA. Reference: PHAK

365. Angle of attack is defined as the angle between the chord line of an airfoil on a fixed wing UA and the

a) direction of the relative wind.

b) pitch angle of an airfoil.

c) rotor plane of rotation.

Answer – a. The chord line of a wing is the imaginary line that dissects the wing in the middle when looking at a cross-section of the wing. The relative wind is the oncoming wind and the angle between the chord line and the wind is AOA. Reference: PHAK

366. Refer to Figure 1: The acute angle A is the angle of

a) incidence.

b) attack.

c) dihedral.

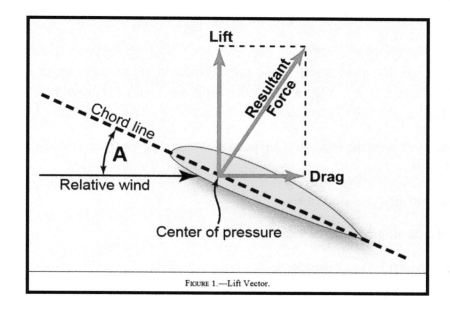

FIGURE 1.—Lift Vector.

Answer – b. AOA is the angle between the oncoming wind, called relative wind, and the chord line of the wing. The chord being the imaginary line that dissects the wing in half. Reference: PHAK

367. The angle between the chord line of an airfoil and the relative wind on a fixed wing UA is known as the angle of

a) lift.

b) attack.

c) incidence.

Answer – b. Angle of Attack (AOA) is the angle between the chord line and the relative wind. Reference: PHAK

368. The angle of attack at which a fixed wing UA stalls, will

a) increase if the CG is moved forward.

b) change with an increase in gross weight.

c) remain the same regardless of gross weight.

Answer – c. An aircraft will always stall at the same critical angle of attack regardless of gross weight. The critical angle of attack is the AOA in which the wing stalls which is around 15° of pitch on average. Reference: PHAK

369. As altitude increases, the indicated airspeed at which a given fixed wing UA stalls in a particular configuration will

a) decrease as the true airspeed decreases.

b) decrease as the true airspeed increases.

c) remain the same regardless of altitude.

Answer – c. The fixed wing UA will stall at the same indicated stall speed (on the airspeed display) regardless of higher true airspeeds in high-density altitude situations. An example would be flight from a high elevation job site in which the aircraft would be at a higher true airspeed, but stall speed would be indicated the same as when at lower altitudes. While there may be a noticeable decline in flight performance. One common error a Person Manipulating the Controls (PMC) may make is intentionally flying faster-indicated airspeeds to avoid stalling. Reference: PHAK

370. In what flight condition must a fixed wing UA be placed in order to enter a spin flight condition?

a) Partially stalled with one wing low.

b) In a steep diving spiral.

c) Stalled.

Answer – c. An aircraft first has to be stalled before it will spin. However, a yawing motion, called uncoordinated flight, must also be subjected to the aircraft for it to spin. Uncoordinated flight is when an inappropriate amount of rudder is added to counteract adverse yaw (the use of rudder to yaw the aircraft when ailerons are used to bank the aircraft). Reference: PHAK

371. During a spin to the left in a fixed wing UA, which wing(s) is/are stalled?

a) Both wings are stalled.

b) Neither wing is stalled.

c) Only the left wing is stalled.

Answer – a. A spin happens when an aircraft is stalled and a yawing motion is subjected to the aircraft (via the rudder). As the aircraft descends it looks as if only one wing is creating lift, but technically both wings are stalled. However, one wing is more stalled than the other. Reference: PHAK

372. How will frost on the wing and propeller(s) of a UA affect the take-off performance?

a) Frost will disrupt the smooth flow of air over the wings and propeller(s), adversely affecting its lifting capability.

b) Frost will change the camber of the wing and propeller(s), increasing its lifting capability.

c) Frost spoils the smooth flow of air over the wing and propeller(s), thereby decreasing lifting capability.

Answer – c. Anything that disrupts the smooth flow of air over a lifting surface disrupts lift. Reference: PHAK, RFH

373. How does frost affect the lifting surfaces of a UA on takeoff?

a) Frost may prevent the UA from becoming airborne at normal takeoff speed.

b) Frost will change the camber of the wing and propeller(s), increasing lift during takeoff.

c) Frost may cause the UA to become airborne with a lower angle of attack at a lower indicated airspeed (UA fixed wing).

Answer – a. Anything that disrupts the smooth flow of air over a lifting surface disrupts lift on takeoff and in-flight. Reference: PHAK, RFH

374. What is ground effect?

a) The result of the interference of the surface of the Earth with the airflow patterns about a UA.

b) The result of an alteration in air flow patterns, increasing induced drag about the wings of a UA.

c) The result of the disruption of the airflow patterns about the wings of a UA to the point where the wings will no longer support the aircraft in flight.

Answer – a. This is a very complex way of saying that ground effect is the cushioning effect of ground when an aircraft is within one half wing span (or one half rotor diameter for rotorcraft). There is also a reduction in induced drag that also promotes additional lift. Pilots should be prepared to handle the effects of ground effect on landing. Reference: PHAK, FAA-H-8083-21, RFH

375. Floating caused by the phenomenon of ground effect will be most realized during an approach to landing in a UA when at

a) less than the length of the wingspan (fixed wing) and one half rotor span (rotor craft) above the surface.

b) twice the length of the wingspan (fixed wing) or rotor span (rotorcraft) above the surface.

c) a higher than normal angle of attack for only fixed wing UA.

Answer – a. Ground effect is the cushioning effect of ground when an aircraft is within one half wing span (or one half rotor diameter for rotorcraft). There is also a reduction in induced drag (drag created in the production of lift) that also promotes additional lift. Pilots should be prepared to handle the effects of ground effect on landing. References: PHAK, RFH

376. What must a Person Manipulating the Controls (PMC) be aware of as a result of ground effect in a UA?

a) Wingtip vortices increase, creating wake turbulence problems for arriving and departing manned aircraft.

b) Induced drag decreases, therefore, any excess speed at the point of flare may cause considerable floating.

c) A full stall landing in a fixed-wing UA will require less up elevator deflection than would a full stall when done free of ground effect.

Answer – b. Pilots should be prepared for the cushioning, or floating, while in ground effect and adapt accordingly in fixed wing and rotorcraft. References: PHAK, Rotorcraft Flying Handbook

377. A fixed wing aircraft leaving ground effect during takeoff will

a) experience a reduction in ground friction and require a slight power reduction.

b) require an increase in angle of attack to maintain the same lift coefficient.

c) require a lower angle of attack to maintain the same lift coefficient.

Answer –b. Once outside of ground effect (more than one-half wingspan or one half rotor span for rotorcraft) lift decreases back to normal levels requiring an increased angle of attack or additional power in rotorcraft. References: PHAK, Rotorcraft Flying Handbook

378. Ground effect is most likely to result in which problem?

a) Setting to the surface abruptly during landing.

b) Becoming airborne before reaching recommended takeoff speed.

c) Inability to get airborne even though airspeed is sufficient for normal takeoff needs.

Answer – b. Due to increased lift and decreased induced drag a pilot may believe the aircraft is ready for takeoff, prematurely. References: PHAK, Rotorcraft Flying Handbook

379.   What force makes a fixed wing UA aircraft turn?

a) The horizontal component of lift.

b) The vertical component of lift.

c) Centrifugal Force.

Answer – a. Looking at a turning aircraft (see graphic) turns are a product of Vertical and Horizontal Lift creating a resultant lift (blue lift arrow), a.k.a. the Horizontal Component of Lift. Reference PHAK.

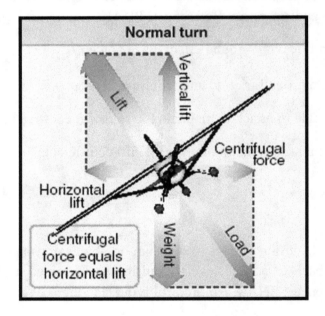

380.   A UA aircraft said to be inherently stable will

a) be difficult to stall.

b) require less effort to control.

c) not spin.

Answer – b. Stable aircraft equals less effort to control, makes sense to me! Reference: PHAK

381. What determines the longitudinal stability of a fixed wing UA aircraft?

a) The location of the CG with respect to the center of lift.

b) The effectiveness of the horizontal stabilizer, rudder, and rudder trim tab.

c) The relationship of thrust and lift to weight and drag.

Answer – a. Longitudinal stability is different that the longitudinal axis of the aircraft. Longitudinal stability refers to the pitch stability. Pitch stability is dependent on the location of the Center of Gravity (CG), or balance point, and the center of lift (or center of pressure). Center of lift (or pressure) is an imaginary point at which the wings lifting force is concentrated. The stability of the aircraft is increased when the center of lift is behind, or aft, of the CG. Reference: PHAK

382. Changes in the Center of Pressure (CP) of a wing on a fixed wing UA affect the aircraft's

a) lift/drag ratio.

b) lifting capability.

c) aerodynamic balance and controllability.

Answer – c. The stability of the aircraft is increased when the center of lift (center of pressure) is behind, or aft, of the CG. As the aircraft angle of attack changes so does the center of pressure, which affects aircraft stability or aerodynamic balance and controllability. Reference: PHAK

383. A fixed wing UA has been loaded in such a manner that the Center of Gravity (CG) is located aft of the aft CG limit. One undesirable flight characteristic a Person Manipulating the Controls (PMC) might experience with this aircraft would be

a) a longer takeoff run.

b) difficulty in recovering from a stalled condition.

c) stalling at higher than normal airspeed.

Answer – b. An aft CG is the most undesirable situation for an aircraft to be in, due to the possible inability to rover from a stall. Reference: PHAK

384. What causes a fixed wing UA (except a T-tail) to pitch nose-down when power is reduced and controls are not adjusted?

a) The CG shifts forward when thrust and drag are reduced.

b) The downwash on the elevators from the propeller slipstream is reduced and elevator effectiveness is reduced.

c) When thrust is reduced to less than weight, lift is also reduced and the wings can no longer support the weight.

Answer – b. With a conventional tail, horizontal stabilizer mid or lower on the empennage (aircraft tail surfaces) the propeller directs air over the horizontal stabilizer creating a natural down-force on tail surface. This is a good design characteristic because when power is reduced airflow is reduced over the horizontal stabilizer and the downforce is less pronounced. When a pilot wants to descend the tail of the aircraft (due to the reduced downforce) is allowed to rise and the nose travels down for the desired descent. Conversely, when the pilot wants to climb, power is increased, the down-force is intensified pushing the tail down and the nose up for a climb. A reduction in elevator effectiveness is a fact, but not a critical factor over the design advantages of a conventional empennage. Reference: PHAK

385. Loading a UA to the most aft Center of Gravity (CG) will cause the aircraft to be

a) less stable at all speeds.

b) less stable at slow speeds, but more stable at high speeds.

c) less stable at high speeds, but more stable at low speeds.

Answer – a. An aft CG is of the least desired conditions because it is less stable at all speeds and can have difficulty in recovering from a stall. Reference: PHAK

386. What is the effect of advancing the throttle in a fixed wing UA while in flight?

a) Both aircraft ground speed and angle of attack will increase.

b) Airspeed will remain relatively constant, but the aircraft will climb.

c) The aircraft will accelerate, which will cause a turn to the right.

Answer – a. Advancing the throttle will enhance the natural down-force on the tail increasing the angle of attack in addition to the groundspeed (if the angle of attack isn't too great or steep). Reference: PHAK

387. What flight condition are torque effects more pronounced in a single engine, fixed wing UA?

a) Low airspeed, high power, high angle of attack.

b) Low airspeed, low power, low angle of attack.

c) High airspeed, high power, high angle of attack.

Answer – a. If you have ever flown an aerobatic RC aircraft, you are familiar with torque rolls in which the torque of the engine makes the aircraft turn about its longitudinal axis. Therefore, low speed, high power and high angle of attack scenarios to enhance the effects of torque. Reference: PHAK

388. The left turning tendency of a fixed wing UA caused by P-factor is the result of the

a) clockwise rotation of the engine and the propeller turning the airplane counterclockwise.

b) propeller blade descending on the right, producing more thrust than the ascending blade on the left.

c) gyroscopic forces applied to the rotating propeller blades acting 90 degrees in advance of the point the force was applied.

Answer – b. P-factor explained: From behind the standard designed fixed wing aircraft visualize the propeller as one large disk that is split in the middle (from the 12 o'clock to 6 o'clock position). As the propeller turns clockwise the descending blade (right-hand side of the disk) creates more thrust than the ascending blade (left-hand side of the disk) making the aircraft want to turn to the left. Reference: PHAK

389. When does P-factor cause the fixed wing UA to yaw to the left?

a) When at low angles of attack.

b) When at high angles of attack.

c) When at high airspeeds.

Answer – b. Like torque, P-factor is the greatest when at high angles of attack. Reference: PHAK

390. The amount of excess load that can be imposed on the wing of a UA fixed wing aircraft depends on the

a) position of the CG.

b) speed of the aircraft.

c) abruptness at which the load is applied.

Answer – b. More speed flowing over the lifting surface (wing) allows more loads to be imposed (handled) by the wing. Reference: PHAK

391. Which basic flight maneuver increases the load factor on an airplane as compared to straight-and-level flight?

a) Climbs.

b) Turns.

c) Stalls

Answer – b. Load factor is the ratio of the lift of an aircraft to its weight (some use the term G-force to understand the concept). During turns, lift is lost, and this loss increases load factor. Reference: PHAK

392. During an approach to a stall, an increased load factor will cause the aircraft to

a) stall at a higher airspeed.

b) have a tendency to spin.

c) be more difficult to control.

Answer – a. An increase in load factor means a decrease in lift. If lift is decreased and the aircraft is carrying more weight it will stall at a higher airspeed. Reference: PHAK

393. Figure 2: If a UA weighs 23 pounds, what approximate weight would the aircraft structure be required to support during a 60 degree banked turn while maintaining altitude?

a) 23 pounds

b) 34 pounds

c) 46 pounds

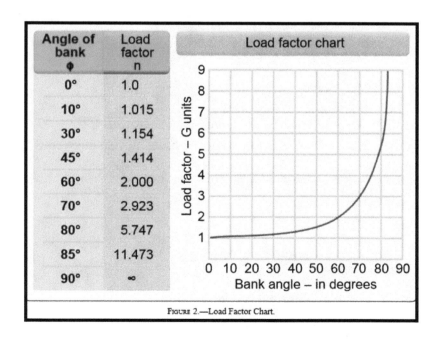

FIGURE 2.—Load Factor Chart.

Answer – c. Find the bank angle of 60° on either side of the chart and see the load factor is 2. Multiply the weight of the aircraft by 2 (23 X 2 = 46). Reference: PHAK

394.   Figure 2: If a UA aircraft weighs 33 pounds, what approximate weight would the airplane structure be required to support during a 30 degree banked turn while maintaining altitude?

a) 12 pounds

b) 31 pounds

c) 38.08 pounds

Answer – c. Find 30° on the left-hand side of the chart for accuracy and you will see a load factor of 1.154. Multiply the weight of the aircraft by the load factor and you get the answer (33 X 1.154 = 38.02 pounds). If your calculator doesn't come up with the same answer, it is OK to go with the closest answer. Reference: PHAK

395.   Figure 2: If a UA aircraft weighs 45 pounds, what approximate weight would the aircraft structure be required to support during a 45 degree banked turn while maintaining altitude?

a) 45 pounds.

b) 63.63 pounds.

c) 72 pounds.

Answer – b. 45 lbs. (aircraft weight) X 1.414 (chart load factor at 45°) = 63.63. If your calculator doesn't come up with the same answer, it is OK to go with the closest answer. Reference: PHAK

396.   Figure 72: A positive load factor of 2 at 80 miles per hour would cause the UA fixed wing aircraft to

a) stall.

b) break apart.

c) operate normally as it is within the normal operating range.

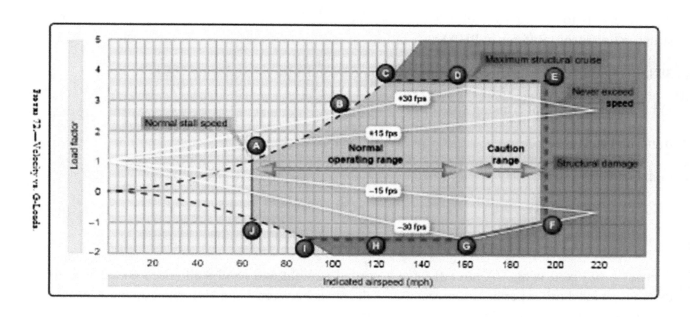

Figure 72 Velocity G-Load graph

Answer – a. While the graphic depicts speeds well over the 100 miles per hour (87 knots) speed restriction for UA (CFR 107.51) the point of the exercise is to consider the effects of load factor with respect to gust, stall speed, and maximum structural airspeed. Reference PHAK

397. Figure 72: What load factor would be created if positive, 15 feet per second gust were encountered at 100 miles per hour?

a) 2.8

b) 3.0

c) 1.75

Answer – c. Find 80 miles per hour on the graph. Travel vertically to the +15 feet per second (fps) gust line. Proceed left horizontally to the load factor. There are four divisions, with alternating colors, between +1 and +2 which mean each block is 0.25 making our answer 1.75. While the graphic depicts speeds well over the 100 miles per hour (87 knots) speed restriction for UA (CFR 107.51) the point of the exercise is to consider the effects of Load Factor with respect to gust, stall speed and maximum structural airspeed. Reference PHAK

398. Figure 72: The airspeed indicated by points A and J for a fixed wing UA is

a) maximum structural cruising speed.

b) normal stall speed.

c) maneuvering speed.

Answer – b. Fixed wing UA aircraft have a normal straight and level stall speed called Vs1. In this example, it is 64 miles per hour represented by the areas between J and H on the graphic. One concept to keep in mind is how a fixed wing UA can fly well below Vs1 in certain situations by the Person Manipulating the Controls (PMC). This is accomplished by using thrust from the propeller exclusively, known as "hanging by the prop" in the aerobatic community. Reference: PHAK

399. Wingtip vortices are created only when an aircraft (manned and UA)

a) operating at high speeds.

b) heavily loaded.

c) developing lift.

Answer – c. Wingtip vortices are created any time a wing is creating lift. The airflow at the wingtip curls up to the top of the wing as the high-pressure underneath seeks the lower pressure on top of the wing. This creates a mini-tornado (or vortex) wind patterns coming from the wind tips. This can be a very dangerous vortex to be caught in manned aircraft, especially when flying behind heavy aircraft on approach to an airport. Reference: PHAK

400. Wingtip vortices created by large manned aircraft tend to

a) sink below the aircraft generating turbulence.

b) rise into the traffic pattern.

c) rise into the takeoff or landing path of a crossing runway.

Answer – a. Wingtip vortices or wake turbulence tends to flow down and away from the aircraft. Resource: PHAK

401. How does the wake turbulence vortex circulate each wingtip?

a) Inward, upward, and around each tip.

b) Inward, upward, and counterclockwise.

c) Outward, upward, and around each tip.

Answer – c. Wingtip vortices or wake turbulence is a function of the high-pressure air under the wing seeking the low-pressure on top of the wing. Resource: PHAK

402. When operating in the airport area where heavy manned aircraft are operating, one should be particularly alert to the hazards of wingtip vortices because this turbulence tends to

a) rise from a crossing runway into the into the takeoff and landing paths of the airport.

b) rise into the traffic pattern area surrounding the airport.

c) sink into the flight path of aircraft operating below the aircraft generating the turbulence.

Answer – c. Heavy aircraft on approach are particularly dangerous because they are traveling at slow speeds and are in a high lift condition, increasing the effect of wingtip vortices. Reference: PHAK

403. The greatest vortex strength occurs when the generating aircraft is

a) light, dirty, and fast.

b) heavy, dirty, and fast.

c) heavy, clean, and slow.

Answer – c. Clean refers to an aircraft that don't have flaps and slats (leading edge wing extenders) extended. Dirty is the opposite in which flaps and slat are deployed. Heavy aircraft on approach are particularly dangerous because they are traveling at slow speeds and are in a high lift condition, increasing the effect of wingtip vortices. Reference: PHAK

404. Which V-speed represents maximum flap extended speed for fixed wing UA that have flaps?

a) V_{FE}

b) V_{LOF}

c) V_{FC}

Answer – a. V-speeds are Velocity speeds which are limitation speeds, such as; maximum landing gear extension, maximum flap operating, etc. V_{FE} is maximum Flap Extension speed. Reference: PHAK

405. Which V-speed represents a maximum landing gear extended speed?

a) V_{LE}

b) V_{LO}

c) V_{FE}

Answer – a. V-speeds are Velocity speeds which are limitation speeds, such as; maximum landing gear extension, maximum flap operating, etc. V_{LE} is the maximum landing gear extended speed or said another way, maximum speed in which the aircraft can be flown with the landing gear in the down (or extended) position. Reference: PHAK

406. V_{NO} is defined as the

a) normal operating range.

b) never-exceed speed.

c) maximum structural cruising speed.

Answer – c. V-speeds are Velocity speeds which are limitation speeds, such as; maximum landing gear extension, maximum flap operating, etc. V_{NO} is the beginning of the "maximum structural cruising speed" in smooth air conditions denoted by the beginning of the yellow arc on an airspeed indicator. Reference: PHAK

407. Which V-speed represents maneuvering speed?

a) V_A

b) V_{LO}

c) V_{NE}

Answer – a. Maneuvering speed is a speed not shown on an airspeed indicator. The function of V_A is to prevent the damage to the aircraft while flying in turbulent weather. If a pilot is at one of these speeds (that vary based on weight) the aircraft wing will stall before damage can occur. The higher weight of the aircraft allows for a higher V_A. Reference: PHAK

408. V_{SO} is defined as the

a) stalling speed or minimum steady flight speed in the landing configuration.

b) stalling speed or minimum steady flight speed in a specified configuration.

c) stalling speed or minimum takeoff safety speed.

Answer – a. Memory aid: This is the stalling speed of the aircraft with Stuff Out (flaps). It is the lowest stalling speed due to the deployment of high lift devices (flaps). Reference: PHAK

409. Which would provide the greatest gain in altitude in the shortest distance during a climb after take-off in a fixed wing UA?

a) V_Y

b) V_A

c) V_X

Answer – c. Vx is a Velocity speed not shown on an airspeed indicator. **Vx** gives the pilot the **greatest gain of altitude over distance**. Vᵧ gives the pilot the **greatest gain of altitude over time**. Pilots will use Vx to take-off from short landing strips and over obstacles. Vᵧ is used for a normal climb. Reference: PHAK

410. After takeoff in a fixed wing UA, which airspeed would the pilot use to gain the most altitude in a given time?

a) V𝗬

b) V𝗫

c) V𝗔

Answer – a. Vx is a Velocity speed not shown on an airspeed indicator. **Vx** gives the pilot the **greatest gain of altitude over distance**. Vᵧ gives the pilot the **greatest gain of altitude over time**. Pilots will use Vx to take-off from short landing strips and over obstacles. Vᵧ is used for a normal climb. Reference: PHAK

411. sUAS operators may get approval to operate on public airport property (called public use airports). What Security Identification Areas (SIDA) security measures should sUAS flight crews be aware of?

a) The Fixed Based Operator (FBO) at the main terminal will let you on the ramp and thus approved for airport property SIDA access.

b) Submit an FAA waiver form to the airport manager for access to the airport property.

c) Many public use airports have Security Identification Display Areas (SIDA) in which airport identification, specific to the airport, is required for all persons inside of secure areas of the airport, or be escorted by a properly credentialed individual. A person can be subject to significant fines if found violating SIDA.

Answer – c. Security is a serious issue at airports in the post 9/11 world. SIDA is the airport badge identification required for access to airport property. The identification is specific to the airport and typically only covers certain areas of the airport. Fines can be significant for not having proper identification. There is a provision to be under the escort of a person with the proper SIDA badge.

Glossary

Above Ground Level (AGL) – Actual height of an object above the ground. For example; a tower that extends from the ground to a height of 200'. Aeronautical charts will express obstacle heights as AGL and above Mean Sea Level (MSL).

Advisory Circular (AC) – A document drafted by the FAA to clarify commonly misunderstood regulations or operating procedures.

Aeronautical Decision Making (ADM) – Closely related to Crew Resource Management as a methodology on how to make effective decisions while utilizing all of the resources available to the flight crew.

Aeronautical Information Manual (AIM) – Formally referred to as the Airman's Information Manual. This publication covers expanded dialog on FAA regulations such as; communications, traffic pattern entries and airspace.

Airmen Certification Representative (ACR) – A person appointed by the FAA to assist in the submission of documents for Remote Pilot (RP) certification.

Airport Facility Directory (AFD) – Also known online as; digital Chart Supplement (d-CS), a directory of airport information such as; hours of operation, services available, elevation, traffic patterns, etc. Information is made available in hard copy or online at; http://www.faa.gov/air_traffic/flight_info/aeronav/digital_products/dafd/.

Airworthiness – When an aircraft is deemed worthy of the air or stated another way means it is in sound mechanical condition for flight.

Air Traffic Control – Personnel responsible for coordinating arriving and departing aircraft.

Air Traffic Control Tower (ATCT) – Facility at the airport in which air traffic controllers coordinate arriving and departing aircraft.

Automated Flight Service Station (AFSS) – Flight weather briefings (in addition to Temporary Flight Restrictions (TFRs) and Notices to Airmen (NOTAMs) available at; https://www.1800wxbrief.com/afss/#!/ or via telephone at 1 (800) WX Brief (992-7433).

Automated Terminal Information Service (ATIS) – ATIS is a lot like AWOS with hourly weather reports for the immediate airport area. The difference being it is used at larger airports and may include additional airport information such as; taxiway closures, communication frequencies, etc.

Automated Weather Observation Station (AWOS) – Automated weather report for the immediate airport area that is issued as a METAR weather report. Closely related to ASOS weather reports, the difference is that ASOS systems don't have the clearways (clearing of trees, buildings, and other obstructions) large enough to qualify for an AWOS.

Beyond Visual Line of Sight (BVLOS) – Flight operations beyond visual line of sight via the use of First Person View (FPV) goggles and Visual Observer (VO) spotters. Not approved unless by FAA waiver.

Certified Flight Instructor (CFI) – A flight instructor authorized by the FAA to provide flight training instruction in manned aircraft.

Challenge and Response (checklist method) – A method of completing a checklist in which verbal commands ensure the items have been completed. For example, the non-flying Remote Pilot in Command (RPIC) will read the first item in a sequence, "Quadcopter power ON" (Command). The Person (pilot) Manipulating the Controls (PMC) will complete the action and respond by saying, "Quadcopter power ON" (Response).

Chart Supplement – Gives more information about an airport than what can be displayed on the aeronautical chart, such as hours of operation, services, contact numbers, etc. Previously known as the Airport Facility Directory (AFD).

Checklist – A list of items placed in a sequence to lessen the chances of skipping important items. This list can be completed by the pilot alone or completed aloud in a crew or team concept. In a team setting a checklist is completed traditionally in a "command and response" format. For example, after starting the propeller motors on a quadcopter, the non-flying Remote Pilot in Command (RPIC) flight team member would read from the checklist, "power developing on all four propellers." After observing that all 4 propellers are developing power the (pilot) Person Manipulating the Controls (PMC) would say, "power on all four propellers."

Code of Federal Regulations (CFR) – Laws or regulations, specifically for our purposes are the CFRs pertaining to UAS operations covered under Part 107.

Common Traffic Advisory Frequency (CTAF) – An aviation band radio frequency used for voice communications at airports without an air traffic control tower. The frequency is used by pilot within 5 to 10 miles of the airport. While the frequency is unique it may be shared by other airports in the region. As a result, good communication technique includes the use of the airport name during transmissions.

Designated Pilot Examiner (DPE) – Independent representatives selected (or designated) by the FAA to administer manned aircraft pilot examinations, called Oral and Practical Test, for airmen certificates and ratings.

Digital Chart Supplement (d-CS) – See Airport Facility Directory (AFD).

Enforcement Action – Action taken by the FAA to enforce adherence to FAA regulations. Similar to a State Highway Patrol stopping a driver for speeding.

Equivalent Level of Safety – A term used by the FAA, typically used on 333 Exemptions, for actions that will provide an equal level of safety in-light of not complying with regulations.

FAA – Federal Aviation Administration, regulates various aspects of the aviation industry from airspace, pilot certifications, flight operations and aircraft certifications. The FAA has field offices called Flight Standards District Offices (FSDO, pronounced, fizz-doe) across the US. Questions related to sUAS operations should be directed to the FSDO has jurisdiction where the flight is taking place. Physical visits to an FAA FSDO office requires an appointment due to security measures.

First Person View (FPV) – Goggles with a visual display of the UA onboard aircraft, used for aircraft control. Not approved unless by FAA waiver.

Fixed Based Operator (FBO) – Facility located at an airport that offers services to aircraft that are transient and based at the airport. For example, services can include; fuel, hangar space, parking, rental cars, maintenance, etc. Note: When notifying a non-ATCT airport regarding a UA flight mission to keep in mind the FBO might be run by a commercial business or the local city or county government or a combination of the two (business runs the FBO and the city manages the airport property and concerns). Two notifications should be sent, one to the FBO business and one to the city, about the flight operation.

Flight Standards District Offices (FSDO, pronounced, fizz-doe) - A FSDO is a field office of the FAA that have locations that cover all regions of the US. Questions related to sUAS flight operations should be directed to the appropriate FSDO having jurisdiction over the flight mission site.

Fuselage – Main body of an aircraft.

Ground School – Text and study guide material used as a self-study or in a course setting to complete an FAA Knowledge Test, for example, the Remote Pilot certificate covered under Part 107.

Integrated Airmen Certification and/or Rating Application (IACRA) – FAA website used for applying for airmen certificates and ratings. Promoted by the FAA to replace less efficient paper application, From 8610-2.

Knowledge Testing Center (KTC) – FAA approved testing centers located across the United States in which persons can take the Knowledge Test (KT), a computer-based exam, for the FAA Remote Pilot (RP) certification in addition to manned pilot KT's.

Line of Sight (LOS) – A sUAS flight crew should always maintain a visual line of sight with their aircraft.

Loss of Control (LOC) – Loss of control due to an equipment malfunction, Person (pilot) Manipulating the Controls (PMC) error, and/or act of nature event, such as a rouge wind gust.

Aviation Routine Weather Reports (Meteorological Aerodrome Report), a.k.a. METARs – Hourly weather reports for airports. Reports can be more frequent, called a Special (SPECI) if there is a significant change in the weather.

Military Training Route (MTR) – Listed on the aeronautical chart as a gray line with associated route type and number. VR routes are Visual Routes and IR routes are Instrument Routes. VR routes are flown visually (eyes outside of the aircraft) and IR routes are flown using the flight instruments inside of the cockpit. If the route has 3 numbers, it is flown above 1,500' AGL. If the route has 4 numbers, it is flown below 1,500' AGL. Aircraft speeds can be in excess of 250 knots.

Mean Sea Level (MSL) – Altitude expression used in aviation to express the height above Mean (average) Sea Level. For manned aircraft altimeters, in the absence of a current barometric setting, pilots are directed to set their aircraft altimeter to the published (Airport Facility Directory – AFD) airport elevation of which they are departing (set prior to departure).

National Airspace System (NAS) – All airspace over US controlled areas in which the FAA is responsible for regulating.

National Security Area (NSA) – Sensitive governmental areas that are no-fly zones due to their national security status.

NTSB – National Transportation Safety Board, handles major aircraft accident investigations. Typically, hands off smaller aircraft accident reporting and investigation to the FAA.

Notices to Airmen (NOTAM) – Additional information that may be of interest to a sUAS flight crew, for example, towers not published on the aeronautical chart, flight restrictions and out of service notices regarding airports.

Operations – Also known as flight. Sometimes referred to as a 1 take-off and 1 landing for airport usage data.

Part 107 – FAA regulations that cover UAS (drone) commercial flying.

Person Manipulating the Controls (PMC) – The person/pilot is responsible for flying the aircraft or the automated flight. However, the Remote Pilot (RP), while not flying, has final authority and responsibility for the safe flight (operation) of the sUAS. The RP and PMC can also be aided by a Visual Observer (VO) for the safety of the flight.

Positive Transfer of Control – A method of changing control of the sUAS from one person to another while on a flight mission. The reason behind this method is to ensure there is always someone in control of the sUAS and that each person understands their role change in the crew. For example, the Person (pilot) Manipulating the Controls (PMC) wants to give his PMC duty to another pilot so he can take a break. As the PMC hands-off the controller to another crew member, he states, "you have the aircraft." The new PMC responds by saying, "I have the aircraft." This allows for no ambiguity regarding the active control of the sUAS.

Pre-Flight – A physical check of the sUAS to ensure it is safe for flight, a.k.a. airworthy. For example, a physical inspection of propeller blades on a quadcopter in addition to other items of importance for flight.

Situational Awareness (SA) – The method in which the sUAS flight crews take in all information related to the flight mission for enhanced safety. For example, the Remote Pilot in Command (RPIC) points out the increase in wind speed between buildings and suggest corrective action to compensate. The suggestion stems from a visual reference, flags, weather forecast and experience. The forecast reported wind levels would be just under the maximum wind amount allowed by the manufacturer for safe flight operations.

Remote Pilot (RP) – The has final authority and responsibility for the safe flight (operation) of the sUAS. He/she may or may not be actively flying the aircraft such as when a Person Manipulating the Controls (PMC) is used in larger flight crew settings. A Visual Observer (VO) may also be used to monitor the safety of the flight.

Security Identification Display Areas (SIDA) – Many airports have identification, specific to the airport, for access to secure areas of the airport. Significant fines can be assessed for violations. There is a provision to be escorted by someone with the proper credentials on airport property. Typically, airport access via SIDA badge is only in specific areas.

Small Unmanned Aircraft System (sUAS) – Also referred to as small Unmanned Aerial System or Drone in the consumer/prosumer aircraft market.

Temporary Flight Restriction (TFR) - A no-fly zone (with very few exceptions granted) that may contain activities such as natural disasters, large events and presidential travel.

United States Code (U.S.C.) – Federal Codes which cover aviation in addition to many more federal laws.

Unmanned Aircraft (UA) – Unmanned Aircraft. Also referred to as sUAS or Drone in the consumer/prosumer aircraft market.

Unmanned Aircraft System (UAS) – Also referred to as Unmanned Aerial System.

Visual Line of Sight (VLOS) – UA aircraft operations are required to be in visual line of sight, not aided by sight enhancement technology (binoculars, video interfaced zoom lens). Operations have to be by the naked eye or regular prescription glasses or contacts.

Visual Observer (VO) – A person that is part of a larger sUAS flight crew that monitors the safety of the flight operation such as to see and avoid other aircraft and hazards. The VO works under direct supervision of the Remote Pilot in Command (RPIC) and the Person Manipulating the Controls (PMC).

Appendix A - Reference Material

Links available at www.RemotePilotAssociation.com/links

*sUAS Remote Pilot Airmen Certification Standards (ACS),

http://www.faa.gov/training_testing/testing/acs/media/uas_acs.pdf.

*Testing Supplement for Private, Recreational and Sport Pilot Knowledge Test,

http://www.faa.gov/training_testing/testing/test_questions/media/sport_rec_private_akts.pdf.

*14 CFR Part 107 (with discussion), http://www.faa.gov/uas/media/rin_2120-aj60_clean_signed.pdf.

*Advisory Circular 107-2, http://www.faa.gov/uas/media/AC_107-2_AFS-1_Signed.pdf.

*FAA Request for Waiver (ATCT Airport Operations request online form and more) –

https://www.faa.gov/uas/request_waiver/.

*Federal Aviation Administration (FAA) – www.faa.gov.

*Knowledge Testing Centers (US)

- http://www.faa.gov/training_testing/testing/media/test_centers.pdf

*FAA Handbooks and Materials,

https://www.faa.gov/regulations_policies/handbooks_manuals/.

* Don't forget to sign up for our newsletter at; **www.RemotePilotAssociation.com** !

Notes:

Notes:

Notes:

Notes:

CPSIA information can be obtained
at www.ICGtesting.com
Printed in the USA
BVHW011918201118
533652BV00006B/32/P

9 780998 128306